Applying the Precautionary Principle

APPLYING the PRECAUTIONARY PRINCIPLE

ADRIAN DEVILLE
AND
RONNIE HARDING
Institute of Environmental Studies
University of New South Wales

The Federation Press
1997

Published in Sydney by
The Federation Press Pty Ltd
PO Box 45, Annandale, NSW, 2038
71 John St, Leichhardt, NSW, 2040
Ph: (02) 9552 2200 Fax: (02) 9552 1681

National Library of Australia Cataloguing-in-Publication

Deville, Adrian
Applying the precautionary principle.

Bibliography.
ISBN 1 86287 203 1.

1. Environmental law — Australia. 2. Environmental policy — Australia — History. 3. Environmental policy — Australia. 4. Conservation of natural resources — Australia. 5. Environmental management — Australia. 6. Sustainable development — Australia — Law and legislation. I. Harding, Ronnie. II. Title.

333.720994

Designed and formatted by St Peters Press, Hornsby Heights, NSW
Printed by Ligare Pty Ltd, Riverwood, NSW

CONTENTS

This project has been assisted by the
New South Wales Government through
the Environmental Trusts

Preface

The Precautionary Principle

Since the early 1990s environmental policy and legislation in Australia has increasingly included the principles of ecologically sustainable development. One such principle is the precautionary principle which requires that we take action to protect the environment in advance of conclusive scientific evidence that harm will occur from some new or continuing human activity.

The principle emerged with no clear 'rules' for its application. Some greeted it as the potential saviour of a rapidly degrading environment, others saw it leading to the cessation of economic growth. The result of its application to date has been neither such extreme and decision-makers in both the public and private sectors are still grappling to understand the principle and to find a solid framework and set of guidelines for its application.

An Environmental Education Trust Grant

The Institute of Environmental Studies at the University of New South Wales was awarded an Environmental Education Trust Grant to prepare a guide to the principle and its application. This book is the result. We are grateful for the grant which enabled Adrian Deville to work at the Institute on preparation of the guide.

Who is the booklet for?

We hope it is useful for a wide range of people: those in government agencies at all levels and in local government who are responsible for applying the principle in the course of their work; those in the private sector, who similarly must take a precautionary approach in preparing development or management plans and in day-to-day operations; tertiary level students who increasingly need to set their technical and professional education in the context of ecologically sustainable development; and those members of community groups or the public who simply wish to understand the nature of the principle, how it may be applied and its potential influence in environmental decision-making.

Help in exploring application of the principle

We were assisted in the groundwork for this booklet by a number of people who participated in three small group workshops on applying the precautionary principle in management of water resources, forests and wastes. We are grateful to the following for their assistance in these workshops but note that they are in no way responsible for the resulting product, any faults in which are entirely ours.

The water resources workshop: Ken Ayers, NSW National Parks and Wildlife Service; Lorraine Cairnes, Fathom Consulting; Carolyn Davies, NSW EPA; Don Geering, Department of Urban Affairs and Planning; Malcolm Hughes, Hawkesbury-Nepean Catchment Management Trust; Judith Jones, Faculty of Law, Australian National University; Tim Macoun, Camp Scott Furphy P/L; Liz Maher, Sydney Water; Christine Mueller, Sydney Water; Guy Russell, Sydney Water; Robyn Tuft, Australian Water and Wastewater Association; Harley Wright, Kimberley-Clark.

The forestry management workshop: Paul Adam, School of Biological Science, University of New South Wales; Anne Conway, Department of Urban Affairs and Planning; David Eckstein, Institute of Environmental Studies, UNSW; Penelope Figgis, Australian Conservation Foundation; Ray Nias, WWF; Jillian Segal, Special Counsel Dunhill Madden Butler and then. Board Member State Forests of NSW; Tim Shepherd, NSW National Parks and Wildlife Service; Bob Smith, State Forests of NSW; Dennis Zines, State Forests of NSW.

The solid waste management workshop: John Denlay, Friends of the Earth; David Eckstein, IES, UNSW; Warwick Giblin, Pacific Waste Management; Gordon McAllister, [Commonwealth] EPA; Neal McQuade, NSW EPA; Stephen Moore, School of Civil and Environmental Engineering, UNSW; Fay Sutton, Nature Conservation Council of NSW.

We are also grateful to a number of others who have assisted either in discussing application of the principle, or in reviewing drafts of the book. Again, they are in no way responsible for the resulting product, any faults in which are entirely ours. They include: Liz Fisher; Alex Gordon; Chloë Mason, NSW EPA; Graham Pyke, Australian Museum; Guy Russell, Sydney Water; Lorraine Phillips, NSW EPA; Yolande Stone, Department of Urban Affairs and Planning; Paul Taylor Department of Land and Water Conservation.

How to use the book

Part 1 of the book provides a brief explanation of the precautionary principle and its emergence in international and domestic legislation and policy. We look at reasons for applying the principle, ask whether it is 'just a new name for common practice', and question whether other institutional changes in environmental management over recent years are not sufficient to ensure a precautionary approach.

Part 2 moves through a sequence of considerations ('steps') which are necessary in applying the precautionary principle. We start by looking at the levels of decision-making at which application of the principle may occur, ranging from operations to strategic policy development. The definition of the principle is dissected to draw out key terms and these are worked through in a series of steps: Step 1 — are precautionary measures needed? That is, are there (i) threats of serious or irreversible environmental damage from a planned or existing activity; and (ii) is there scientific uncertainty regarding the outcomes. These two criteria are combined to answer the question as to whether there is threat of *significant* (serious + irreversible) damage to the environment.

Step 2 considers 'how precautious should we be?' That is, consider the level of significance of threat and degree of uncertainty over outcomes and decide how much precaution is warranted.

Step 3 surveys the types of precautionary measures that could be applied and the levels of decision-making at which these are located.

Step 4 puts Steps 2 and 3 together by weighing up the types of precautionary measures available and their costs, and the extent of precaution required. Options are developed and assessed.

We believe that working through these steps provides a systematic framework for application of the precautionary principle. This cannot be an entirely objective exercise, but following such steps puts some rigour into an otherwise very open ended decision-making process.

Part 3 consists of 6 tables outlining a range of possible precautionary measures. These tables correspond to the 6 levels of decision-making described earlier in the book. These can be read alone or as a follow up to working through the Steps. Tables 1-6 are an important complement to Step 3.

Feedback

We hope that you find the book useful and would be grateful to hear of any experiences you have had in applying the precautionary principle. You can contact me on 02 9385 4973, fax 02 9663 1015 or E.mail: R.Harding@unsw.edu.au

THE UNIVERSITY OF NEW SOUTH WALES

Ronnie Harding
Director, Institute of Environmental Studies,
University of New South Wales
Sydney 2052
NSW Australia

INSTITUTE OF ENVIRONMENTAL STUDIES

Prologue

◆ The Australian Fisheries Management Authority limits access to new areas for fishing and to development of new fisheries, *pending scientific assessment* of the long-term sustainability of the resources in question.

◆ In 1995 the Australian Cabinet endorsed 'Deferred Forest Assessments'. These temporarily set aside areas of forest from logging *pending information as to the conservation significance of these areas.*

◆ In 1991 a construction company in Denmark received approval to build the Great Belt bridge — the largest bridge in Europe — *but under strict conditions.* The area was already prone to eutrophication and there was concern that dredging may release nutrients. Chlorophyll (as an indicator of eutrophication) was to be monitored continuously at a number of sites by automatic recorders. A rise in chlorophyll over a given amount automatically triggered a feedback mechanism in the management system. This involved initiation of a wide scale survey by consultants, without the need to obtain permission of the construction company. If the algal bloom was localised to the construction area dredging had to stop until the bloom subsided. This meant halting the activity of the world's largest dredger costing tens of thousands of dollars a day. Numerous other feedback loops, all triggering responses, were also in place. These involved oxygen depletion and impacts on duck populations and herring spawning.

◆ In Australia, the Great Barrier Reef Marine Park Authority requires a *performance bond* to be lodged by those who obtain a permit to install a facility on either land or water within the Park. This bond can be used for remedial work if developers fail to meet their obligations regarding impact on the Park arising from their facility.

◆ The International Maritime Authority in 1995 drafted guidelines for activities of its members which *may* have an impact on the environment. One guideline required that 'in developing environmental control measures, consideration should also be given to the steps needed for their effective implementation. This might include enforcement provisions, verification procedures and techniques, achievable schedules and deadlines, and capacity building and technical co-operation programs for developing countries'.

◆ In a 1993 decision in the Land and Environment Court of New South Wales, planning permission for a road through native forest was set aside on the grounds that by destroying habitat for an endangered species it *might* have brought it closer to extinction.

◆ In April 1996 a Commission of Inquiry found that likely environmental impacts from a $1.2 billion gold mine at Lake Cowal adequately met relevant planning and environmental guidelines. Despite this the Government rejected the proposal on the grounds that 'the *unknown* risks to this significant environment [National Estate-listed wetland] can only be avoided by refusing the proposal'.

These actions and decisions all have something in common. Action in each case is taken on the grounds that serious or irreversible harm might occur to the environment. Importantly this action was taken in advance of conclusive scientific evidence that such harm would occur. This is basically the precautionary principle.

PART ONE

INTRODUCING THE PRECAUTIONARY PRINCIPLE

Definition of the Precautionary Principle

Where there are threats of serious or irreversible environmental damage, lack of full scientific certainty should not be used as a reason for postponing measures to prevent environmental degradation. In the application of the precautionary principle, public and private decisions should be guided by:

(i) careful evaluation to avoid, wherever practicable, serious or irreversible damage to the environment; and,
(ii) an assessment of the risk-weighted consequences of various options.

The Intergovernmental Agreement on the Environment (May, 1992)

> Under the precautionary principle it is the environment rather than those whose actions may impact on it, that is given the benefit of doubt.
>
> In contrast, past practice has generally been to require proof of damage before taking actions to avoid *possible* damage.
>
> Indeed, uncertainty regarding harmful effects has often been used as a reason against taking protective measures.

- What *exactly* does this mean for decision-making which may impact on the environment?
- How might it influence your activities?
- What is the background of this principle and why has it been put in place?
- How important is it?
- Who is required to apply it?
- Is it really a new concept, or have we always been 'precautious'?

The Emergence of the Precautionary Principle

At the Second International North Sea Conference in Europe in 1987 the participating countries agreed to control the emissions of substances into the North Sea which may be harmful to living organisms. This agreement was to apply not only to those substances which were already known to be harmful but also to substances which might be harmful but for which *no clear scientific evidence of a causal link between emissions and effects was available.* Such substances were deemed to be those that are persistent, toxic and likely to accumulate in living tissue.

This requirement, referred to as 'the principle of precautionary action', is the first explicit inclusion of the precautionary principle in an international agreement.

Realisation that detrimental changes had occurred to North Sea ecosystems since the 1930s, even though these could not necessarily be linked to specific discharges, was the reason for this action. The long-held assumption that the vastness of the North Sea would assimilate discharges without effect came under question. Instead, it was deemed prudent to regard particular substances as harmful until good evidence to the contrary was available.

This action by nations around the North Sea has led to questioning the basis of responsibility in environmental decision-making.

The precautionary principle requires that:

- we focus on *lack of evidence* for making environmental management decisions;
- we *consider* actions to avoid damage which appears *possible* but not certain.

Under the precautionary principle the 'onus of proof' regarding impacts has shifted to those whose actions may cause change. They are required to provide a convincing argument that their actions will not have serious or irreversible impacts on the environment exceeding long-term benefits.

This shift in onus has the potential to revolutionise environmental management. For some this represents a profound threat to future human progress since it is seen as stalling 'development'. For others it offers the hope that through this changed ethos, we will have a chance of achieving ecologically sustainable development (ESD).

Discussion about the meaning of the precautionary principle has been complex, generally theoretical and confusing. We are only just beginning to explore the possible implications of this shift in responsibility in terms of practical application. This booklet is a contribution to this exploration. It aims to provide practical guidance to all whose decisions and activities may involve application of the precautionary principle. As we shall see below, this means that its target audience is very wide — we are all *potentially* included in this group, both in our working lives and as environmentally responsible citizens. The application of the precautionary principle is not confined to environmental management professionals.

Who is Involved?

Not only designers, managers, developers and operators, but also organisational managers, administrators, legislators, strategic planners and policy makers in all sectors, should be involved in applying the principle (see Figure 1, Part 2).

Whereas the principle was originally applied to marine pollution, it is now considered in relation to a wide range of activities. These may include alteration of ecosystems by clearing land, changing hydrological regimes through damming or altering river courses and modifying fire regimes through land management practices. Even attempts to 'improve ecosystems' will require precautionary considerations. Increasing exposure of humans and other organisms to physical agents such as electromagnetic radiation (e.g. from high voltage transmission lines), noise and seismic vibrations (e.g. seismic waves used in oil exploration) also requires consideration of precaution.

Because the precautionary principle potentially applies to this wide range of activities it affects decision-making across all sectors:

- primary — mining, agriculture, fisheries and forestry;
- secondary — processing of raw materials and manufacturing;
- tertiary — the service sector both public and private. Financial services, tourism, government administration and planning, transport, infrastructure development, consulting.

Those in research and education also play a role through researching techniques for application, teaching the concept of precaution and reducing uncertainty by improving our knowledge.

Why Apply the Precautionary Principle?

Following its inclusion in the North Sea agreements, the principle has been incorporated into numerous international treaties and conventions relating to the environment and is regarded as one of the key principles of 'sustainable development'. From this broad international base it has been included in *national* legislation and policies concerning environmental management. Australia is no exception.

As a result, there are now many reasons why the principle should be applied in decisions which may affect the environment, including policy, law, economics and moral/ethical considerations.

Policy

Two major policy documents in Australia which include the precautionary principle as a guiding principle for environmental management are:

- The National Strategy for Ecologically Sustainable Development (NSESD) (1992);
- The Intergovernmental Agreement on the Environment (IGAE) (1992).

These documents are broad *national* agreements regarding environmental management and the IGAE was signed by representatives of all levels of government in Australia.

Although they are policy documents and not legally binding, they have greatly influenced the incorporation of the principle into environmental legislation, plans, regulations and policies at all levels of government. In turn professional organisations, such as the Institution of Engineers, have included the principles of ESD into their Codes of Practice and have started to explore application of the principles in the context of professional practice. Both public and private organisations have begun to include principles of sustainability, such as the precautionary principle, into corporate plans, mission statements and organisational strategies.

Box 1 — Ecologically Sustainable Development

Sustainable development has been defined as:

> 'development that meets the needs of the present without compromising the ability of future generations to meet their needs' (WCED 1987).

This concept received worldwide attention following the publication of *Our Common Future* in 1987 (WCED 1987) and the United Nations Conference on Environment and Development in Rio de Janeiro in 1992.

The definition is very open-ended giving no clear guidance for application. 'Principles of sustainable development' include intergenerational equity, intragenerational equity, conservation of biological diversity, the precautionary principle and the internalisation of environmental costs.

In Australia, the term ecologically sustainable development (ESD)* has been typically used and a major discussion of ESD took place in the early 1990s resulting in the development of Australia's *National Strategy for Ecologically Sustainable Development* (1992).

A key point in the National Strategy is that the principles of ESD should be treated as a 'package' — no one principle is more important than any other — each should be applied with consideration to the other principles.

*For discussion of ESD see: WCED (1987); Commonwealth of Australia (1992): Beder (1996)

The IGAE requires both 'public and private decisions' to be guided by the precautionary principle. As a government agreement the IGAE only affects government decision-makers *directly*. But it affects the decisions of the private sector because it is now part of the broader policy context for development. Non-government decision-makers may be affected by the precautionary principle wherever their activities are approved or regulated by government, *in addition to the regulations already enforced by specific laws.*

Law

Almost all recent *international treaties* that relate to protection of the environment, include the precautionary principle. These cover fishing resources, climate change, biodiversity, marine pollution and hazardous wastes — and the list grows steadily.

Many of these influence policy and legislation within Australia. Typically, national policy or legislation will follow an international agreement and in turn may lead to complementary legislation at State-level and to development of industry standards. The United Nations Framework Convention on Climate Change (concerning the Enhanced Greenhouse Effect) is a good example. Australia signed this treaty in 1992 and has responded to its commitment by:

- undertaking the National Greenhouse Gas Inventory (1994);
- producing the National Greenhouse Report (1994) to show how Australia is meeting its obligations under the Treaty;
- releasing *Greenhouse 21C* (1995) and developing the Greenhouse Challenge involving a program of co-operative agreements with industry for reductions in greenhouse gas emissions.

In turn, State and local governments have also introduced strategies for addressing the Enhanced Greenhouse Effect.

The precautionary principle has been included

in Australian legislation either by specific reference, by reference to clauses of the IGAE which contain the principle or by general reference to the principles of ESD. Typically, since the early 1990s, new environment-related legislation or amendments to previous legislation have included the precautionary principle through one of these means. Some examples of such legislation are:

- *Protection of the Environment Administration Act* 1991 (NSW)
- *Environment Protection Act* 1993 (SA)
- *National Environment Protection Council Act* 1994 (Cth)
- *Environmental Planning and Assessment Act* 1979 (NSW) 1994 Regulation.
- *Environmental Protection Act* 1994 (Qld)
- *Fisheries Management Act* 1994 (NSW)
- *Water Board (Corporatisation) Act* 1994 (NSW)
- *Environmental Management and Pollution Control Act* 1994 (Tas)
- *Threatened Species Conservation Act* 1995 (NSW)
- *Living Marine Resources Management Act* 1995 (Tas)
- *State Owned Corporations Amendment Act* 1995 (NSW)
- *Energy Services Corporation Act* 1995 (NSW)
- *Electricity Legislation Administration Act* 1995 (NSW).

The *Fisheries Management Act* 1994 (NSW) is an interesting example of the application of the precautionary principle. It not only requires precaution to be applied in setting the total allowable catch, but also performance indicators must be included in management plans for share management fisheries to monitor whether the objectives of the plan and the principles of ESD are being met.

Economics

Micro economic level

At the level of the firm or government agency, application of the precautionary principle is increasingly likely to make good economic sense. This is true from the formulation of plans for a development right through construction and operations to post-activity closure. It is also true in relation to products and technologies.

Developing plans People in industry know it can be very expensive to prepare development plans and applications involving environmental impact assessments. Increasingly, in accordance with legislation, government agencies and third parties (with rights under legislation) will require precaution to have been applied in development of the proposal (see the NSW Department of Urban Affairs and Planning draft guidelines on the 'Practical Application of the Principles of ESD in EIA', 1995). If this is not done developers are likely to be sent back to the 'drawing board', hence multiplying their expenses.

Liability and compensation Failure to apply appropriate precautionary measures during operations and in product development may see a corporation and its senior officers facing heavy compensation cases or even imprisonment. Corporations will find it important to apply precautionary measures as one way of demonstrating 'due diligence'.

Corporate image Environmental responsibility is increasingly seen as part of a positive corporate image, which may bring important economic spin-offs. A precautionary approach to all its activities that may have an environmental impact is today a key component of environmental responsibility for a corporation.

Increasingly it makes good economic and financial planning sense to make sure that the precautionary principle has been used to guide private sector planning and development decisions in their early stages.

Macro economic level

The precautionary principle aims at ensuring that long-term costs do not exceed short-term benefits of an activity. We should not impose burdens of environmental degradation on future generations. Nor do we want to impose costs of environmental degradation on ourselves later in life!

Applied *appropriately*, the principle makes good economic sense for society at large and in terms of equity between generations. However, achieving this economic balance will not be easy. By definition, the application of precaution means that we are dealing with situations that are rife with uncertainty! This makes it difficult to know the optimum level of precaution in economic terms — *too much* precaution may add unnecessarily to today's costs and deny important social benefits.

The difficulty is compounded because 'valuing' components of the environment (such as loss of a species) in order to assess costs and benefits, is not straightforward. As well, the costs and benefits of activities are often borne by different goups of people, both at the present time and between this and furture generations. This imposes further complexity in any analysis.

Moral and Ethical Considerations

As one of the principles of ESD, the precautionary principle underpins the 'new ethos' that guides us into the 21st century. It has importance in terms of encouraging people to be responsible toward one another and, most importantly, to be responsible towards those not yet born. Together with other 'principles of sustainability' (such as conservation of biodiversity) it also has relevance for our responsibilities to all the non-human inhabitants of the planet.

A New Name for Common Practice?

Some have queried the attention being given to the precautionary principle, suggesting that it is 'nothing new' and that we have 'acted with precaution' in many situations long before mention of the principle in environmental policy and legislation. Certainly many such examples can be found, but equally, there are many examples of non-precautionary behaviour (see Box 2).

While we can certainly find examples of the previous application of precaution, these have been ad hoc and have not rested on a formal and consistent requirement for such an approach.

Perhaps what is most significant is that the precautionary principle has now been 'institutionalised' in environmental decision-making and in many cases its consideration has specific legislative backing and is supported by case law.

There have been many other institutional changes in environmental management in recent years. Aren't these enough to ensure a precautionary approach?

Isn't Risk Assessment Enough?

'Risk' refers to situations where we are fairly well-informed about threats. The behaviour of the system under investigation, for example a chemical process plant, is basically well-known and the chances of different outcomes can be defined and quantified by structured analysis of mechanisms and probabilities. By definition the precautionary principle deals with situations where we are much less certain of outcomes.

Hence risk assessment only deals with a sub-set of all the situations in which we should act to prevent serious or irreversible environmental damage. It is not a substitute for a precautionary approach. Nevertheless, a precautionary approach can be taken in risk assessment.

Isn't Environmental Impact Assessment Enough?

Not necessarily.

Uncertainty is not emphasised in EIA

The emphasis in EIA is on predicting impacts that we have some knowledge about. There is far less attention given to highlighting areas where our understanding is poor, where there is great scope for ignorance or where the confidence level attached to scientific assessments is low. Indeed it may not be possible to say with any scientific certainty what the impacts of a proposal will be.

Box 2 — Non-Precautionary Actions

Acid rain

For many years Scandinavian countries maintained that sulphur dioxide emission from British power plants and industry was the cause of acidification of Scandinavian lakes. The British used lack of direct cause-effect evidence as a reason for not taking costly action to reduce these acid emissions. Under a precautionary approach, the shifting of the onus of proof would have put pressure on the British to demonstrate that their emissions were not a primary cause of the acidification of the Scandinavian lakes.

Pesticides

Within Australia numerous groups have lobbied for many years against the widespread use of a range of pesticides in agriculture and property protection, on the grounds that these were harming the health of humans and other species. Until relatively recently, the onus has been on the complainants to demonstrate direct cause-effect links. Of course this is very difficult when dealing with organisms which are part of a complex web of relationships within an ecosystem and which are exposed to many substances. It is especially difficult when the effects may be delayed, as for example with human carcinogens where there may be 30 years or more between exposure to the carcinogenic substance and manifestation of the cancerous effect. Today, we need to feel confident that there will be no harmful long-term effects before allowing use of such substances, rather than the other way around. In other words, the 'burden of proof' has been shifted. As a result, a number of pesticides that have been used for many years for activities such as preventing termites in buildings, have been banned for this use.

Asbestos

It is now well accepted that asbestos causes the potentially fatal diseases of mesothelioma, lung cancer and asbestosis. The time between exposure and effect is delayed in some cases by up to 30 years or more, making cause-effect links difficult to prove conclusively. There were, however, clear signs from the 1930s of the harmful effects of asbestos and a North American insurance company back in 1917 had the commercial 'wisdom' not to insure asbestos workers. Nevertheless, it took until the early 1980s before asbestos was banned for most uses in Australia. It was not until authorities had strong epidemiological evidence that they took action, despite suspicious, but not conclusive, evidence available as early as the 1930s. A consequence of the lack of precautionary action is that many people have died from asbestos-related disease, there have been large claims against corporations involved with asbestos mining and manufacture, and Australia has the highest rate of mesothelioma in the world.

Until fairly recently, acidification of northern New South Wales rivers was not a predicted impact of developments in the region involving disturbance of soils. This is despite much earlier *suggestions* by scientists of the potential for acidification. This is exactly the value and the purpose of the precautionary principle. It forces us to give due consideration to the *possible* impacts that we *aren't certain about* as well as those for which our evidence is good.

The requirement for EIA is narrow

EIA currently only applies to particular kinds of 'projects'. There are a number of other sorts of decisions that could be guided by the precautionary principle that EIAs do not address, such as developing and administering policies, strategies and procedures. So just because many activities of an organisation might require EIA before they can proceed, there may be other decisions that require a precautionary approach that are overlooked by the EIA process.

The scope of EIA is narrow

Assessment of projects in New South Wales under the *Environmental Planning and Assessment Act* 1979 (NSW) requires cumulative impacts to be considered (that is, the additive and synergistic impacts). However, this rarely occurs since in practice, how it should be done and what responses a decision-maker might make, are far from clear. Also, monitoring and feedback on the environmental effects *after* the development is in place are typically not a condition of approval. This means that approved projects may cause serious or irreversible damage that was not predicted in the original assessment, but which may have been picked up at an early stage if monitoring was a requirement.

EIA cannot substitute for the precautionary principle

An important avenue for applying the precautionary principle is certainly to assess intended activities very carefully for their impacts on the environment. However, because of the limits to EIA as outlined above, it cannot be a panacea for preventing environmental damage. It should be thought of as one tool feeding into a process that ultimately ends with a political decision.

Identification

Identification of the capabilities and constraints of land for various forms of development provides the foundation for sustainable land uses. The preparation of Environmental Planning Instruments (land-use plans) is required under the *Environmental Planning and Assessment Act* for this purpose. Environmental Impact Assessment, even if carried out with an emphasis on precaution, cannot make up for a precautionary approach to the development of planning instruments which provide the guidelines for land use.

EIA can become more precautionary

A precautionary approach will be increasingly required as part of EIA as the principles of ESD are included in the relevant legislation. For example, a 1994 Regulation to the *Environmental Planning and Assessment Act,* has required consideration of the principles of ESD in the preparation of an Environmental Impact Statement (EIS).

Isn't an Environmental Management System Enough?

It might be. Certainly development and implementation of an environmental management system (EMS) for an organisation, *in itself,* may be considered as contributing to a precautionary approach.

EMSs set in place 'quality' organisational structures and processes for meeting environmental policy objectives. Not all EMSs have the precautionary principle 'built in' to their framework or to the policy they seek to implement. In such cases there is no guarantee of application of the precautionary principle by the organisation, even though the EMS may lead to overall improvement in environmental performance. Table 3 shows integration of a precautionary approach within an organisation through the development of an EMS.

Isn't Staying Within the Regulations Enough?

Some of your activities probably have to comply with environmental regulations. If regulations are drafted on the basis of a precautionary approach, by complying with these regulations you will be applying precaution. Compliance may involve a licence to operate in a particular area, meeting standards for discharges during operations, or use of the best available technology to carry out your activities.

However:

- not all of your activities may be governed by regulations;
- your activities may be governed by regulations but these may not be based on a precautionary approach;
- the regulations may have been based on a precautionary approach, but conditions may have changed and the regulations may not have been appropriately updated. For example:
  - — new information may become available necessitating stricter emission standards as a precautionary response, or
  - — the regulations may have been based on understanding of impacts in one area, but activities may be taking place in another area for which we lack necessary information.

So while regulations can be a precautionary measure, they may not be enough to ensure application of the precautionary principle in relation to your range of activities.

You should not judge whether a decision requires a precautionary approach on the basis of whether it is already regulated.

Applying the precautionary principle means being proactive. This means being on the look-out for potentially damaging actions before they occur. So it should be applied in addition to existing regulations wherever activities pose a potential threat of serious or irreversible environmental damage.

Making Sense of the Precautionary Principle

The precautionary principle is stated in terms which are open to interpretation. For example: What is a '*serious* threat'? Isn't there *always* some 'scientific uncertainty'? What kinds of 'measures' are *appropriate*? Responses will be influenced

Box 3 - The ISO 14000 Series for EMS

The concept of environmental management systems (EMS) has moved quite rapidly from principles into practice in the 1990s. The adoption of EMS has been formalised through the development of the International Standards Organisation's (ISO) 14000 series of standards for Environmental Management Systems.

Australian companies are now being accredited for these standards which, as well as providing general guidelines on principles and EMS techniques, also cover such environmental management fields as life-cycle assessment, environmental auditing and certification for environmental auditors.

by individual values, preferences and judgements.

Hence the principle is only a *guiding* principle. It is not a hard and fast rule leading to a single 'answer' for application in a given situation. This makes it challenging to apply.

Those charged with implementing the principle want a *consistent* set of rules for knowing when precaution is required and how it should be applied. The lack of clear rules causes decision-making involving the precautionary principle to be an uncertain business. This has led to concern regarding the unpredictability of interpretation and the potential impact on investment. It is quite likely that even in similar developments different decision-makers and communities will reach different decisions regarding application of precaution:

- Will companies be prepared to invest in development plans unless they can be confident that their interpretation of an appropriate level of precaution matches that of the decision-makers and community stakeholders?
- Can these views be determined in the early stages of planning?

Some fear a very strongly precautionary interpretation of the principle that will lead to a stalling of development. Others are concerned that the difficulty in applying the principle will lead it to be ignored by decision-makers.

Finding ways to assist in the application of precaution in a *consistent manner* will have both environmental and economic benefits. Working through a sequence of steps which address the key terms in the IGAE definition of the precautionary principle will lead to more consistent application of the principle. These steps are set out in detail in Part 2.

PART TWO

APPLYING THE PRECAUTIONARY PRINCIPLE

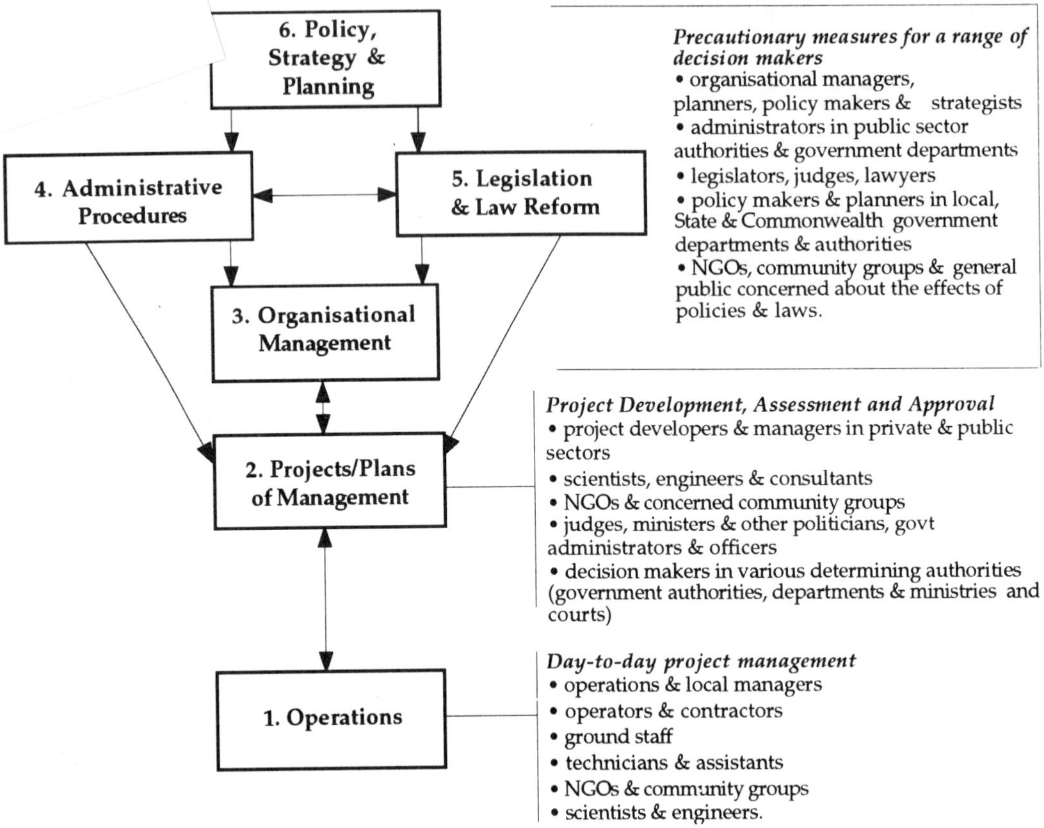

Figure 1: Different precautionary measures and levels of decision-making

Is the Precautionary Principle Relevant to Your Activities?

Yes. All activities and decisions should be *scrutinised* for their potential environmental impact. At one end of the spectrum of possibilities, a householder may consider alternatives in a domestic situation, in which case the potential application of precaution will rest on grounds of both commonsense and the 'ethos of sustainability'; there will be no formal requirement. At the other end of the scale, there may be formal legislative requirements for application of precaution in, say, Environmental Impact Assessment or through the terms of reference for a Public Inquiry.

Simply considering, in a structured manner, the potential for threats of environmental damage, is the first stage in application of the precautionary principle.

But of course, we need to do more than just consider the *potential for threats*. We need to consider whether precautionary measures *are* required and if so *how much* precaution and what *type of measures* should be put in place. Many people at many different levels of decision-making and application are involved.

They include people involved in *operations* — the waste disposal contractor, the logger, the farmer, the chemical plant operator — through to those involved in *strategic planning* and the Chief Executive Officers of organisations. The range of levels of decision-making at which precautionary measures should be considered and/or applied is shown in Figure 1.

In Part 2 we work through the set of questions just outlined, as a sequence of 'Steps':

- Are precautionary measures needed? (Step 1)
- What precautionary measures *can* be applied? (Step 2)
- *How* precautious should we be? (Step 3)
- What precautionary measures *should* be applied? (Step 4)

Each of these steps consists of a range of suggestions, information and questions. For Step 2, greater detail about possible precautionary measures is given in the six Tables in Part 3. These tables cover the six levels of decision-making and application as shown in Figure 1.

Step 1: Are Precautionary Measures Needed?

Precautionary measures are needed when there are *both*:

1 threats of serious or irreversible environmental damage; *and*,
2 lack of scientific certainty about these threats.

In some situations it is obvious that there are serious or irreversible threats *and* that uncertainty about these exists (see Boxes 4 and 5). In other situations it is not so easy to judge. The situation may be very complex or the possible threats from a particular action difficult to predict. Others may alert you to the fact that the precautionary principle is relevant to a particular activity because you have not recognised the potential seriousness or the irreversibility of a situation.

Judging how serious environmental damage might be, or how certain the possible effects are, is open to interpretation. Different people have different opinions!

Are Threats Serious or Irreversible?

People hold different values or 'world-views'. For example, some regard the Antarctic region as the 'last great wilderness' which should be protected, yet others describe it as an abundant 'source of resources' to be exploited. There is also a range of perspectives in between.

Judging the *seriousness* of environmental damage depends on a number of factors. For example, if a dam is constructed for the generation of electricity, the seriousness of potential threats depends on:

- the different values placed on the existing river and surrounding ecosystem by the general public, scientists, water resource administrators, downstream irrigators and other interested parties *(what is the degree of 'acceptable' change?)*;
- the 'boundaries' for defining and assessing the ecosystem and the possible threat *(what is relevant to the assessment?)*;
- the completeness of knowledge about the area *(e.g. are there endangered species?)*;
- the care taken in construction, the scale of the dam and the ecological effects of the dam *(size, spatial scale and complexity of environmental impacts)*.

Judging the *reversibility* of environmental damage is also a subjective matter because:

- there may be scientific disagreement about the ability of the ecosystem to regenerate after construction, or the time needed for this regeneration (remembering of course, that some things such as species loss, are *clearly* irreversible); and,

- there may be different assumptions about the resources available or needed to restore the area.

Determining whether there will be serious or irreversible environmental damage involves social as well as scientific and economic considerations — people's perceptions and values are involved.

How to Reach Consensus?

When assessing the seriousness or irreversibility of a threat it is necessary to:

1 ***Involve a range of views in assessing whether threats are serious or irreversible.***

Broadening the range of professional expertise and strengthening community involvement will help to achieve a consensus about this.

2 ***Try to define types of threats that most people would see as serious or irreversible.***

Scientific opinion and consideration of the full set of 'sustainability principles' suggest the following:

- ***Loss of species*** — at the present time and for the foreseeable future 'extinction is forever'; that is, it is irreversible. Seriousness of loss may vary between species.
- ***Loss of biodiversity*** — biodiversity comprises not just 'species' diversity but also genetic and ecosystem diversity.
- ***Damage to ecological processes*** — breakdown mechanisms leading to cycling of materials through the terrestrial, atmospheric and aqueous sectors of the earth. Loss of such processes, which may be closely tied to biodiversity loss, amounts to loss of 'free essential planetary services'.
- ***Contamination of soils, water bodies and food chains*** with harmful substances that are pervasive in ecosystems either because they bioaccumulate (build up in living tissue) or because they have a long half-life (the period until they break down into possibly less harmful substances is very lengthy). These may include toxic, radioactive, mutagenic (causing mutations or genetic change) and cancer causing substances.
- ***Introduction of 'exotic' organisms to ecosystems*** — these may include species which are not found in the ecosystem in question or they may include 'new' genetically modified organisms. In either case we must, in the absence of good evidence to the contrary, suspect that ecological disturbance will result.
- ***Release of 'new' chemicals*** — many new chemicals are developed each year. In the absence of good evidence that these will not cause harm in the situations in question, we need to be suspicious about the effects of such chemicals which have not previously existed in the Earth's ecosystems.

There may be other 'threats' you think should be added to this list.

Focus on defining *all* aspects of threats associated with your activity and then assess their significance bearing in mind the generic list above and by considering the factors below. Rank your responses on a scale. This will help you to structure the assessment of whether potential threats are significant ('serious' or 'irreversible'). If this process is carried out with a representative set of stakeholders, it can assist in putting some rigour into an otherwise fairly subjective process.

3 ***Consider all aspects of the threat to determine its overall significance***

Numerous factors make threats 'serious' or 'irreversible'. Consider each of these aspects separately and rank your answer on a scale such as those given below.

(a) *What is the spatial scale of the threat?* Environmental effects of an action/activity may be on a small or localised scale or may have broader impacts — regional, statewide, national or global. For example, the management of solid wastes may have local or regional effects on the environment, whereas the release of greenhouse gases has global impact.

❑ local ❑ regional ❑ statewide ❑ national ❑ other

Box 4 - Threats to the Environment

The types of environmental threats the precautionary principle aims to avoid are many. The following are some examples (possible causes are in parenthesis):

- global warming (excessive discharges of CO_2, methane);
- biodiversity loss (land/habitat clearing, demand and markets for products from endangered species, the release and establishment of non-native plant and animal species);
- stratospheric ozone depletion (CFCs, HCFCs)
- desertification and land degradation (land clearing for unsustainable agricultural practices);
- marine ecosystem health (spills and dumping of pollutants);
- freshwater ecosystem health (discharge of pollutants, development activities in riparian zones, recreational activities in waterways);
- atmospheric pollutants (lead, acid rain);
- damage to the structure and function of specific ecosystems and 'renewable resource' bases (overfishing, overstocking of land); and,
- damage to human and non-human physical and mental health (lead, chemical residues in food).

These threats to the environment come from a range of activities:

- *Obtaining resources* — either extracting non-renewable minerals and energy or harvesting and managing 'renewable' resources such as fish or forest timbers.
- *Transforming or using these resources* — constructing buildings, bridges and other infrastructure, manufacturing products, or burning fossil fuels.
- *Disposing of unusable 'by-products'* — managing, reusing, recycling or disposing of waste materials from obtaining and transforming resources.

Threats to the environment may be direct or indirect:

Direct

Physical disturbance to the environment through mining, agriculture, logging, construction, dredging waterways, damming rivers altering the physical configuration of ecosystems. Other physical agents include heat, radioactivity, noise or other forms of electromagnetic energy.

Chemical contamination of the environment through accidental or deliberate discharge of toxic substances to land or water, discharge of gaseous pollutants.

Ecological disturbance either as a result of physical or chemical impacts on the environment, or by introduced species or direct disturbance by humans.

Indirect

Government policies and strategies at a range of levels including planning decisions, changes to administrative procedures, taxes and charges, development of regulations and legislation.

Organisational policies and strategies in corporations, public agencies and other organisations — strategic, financial, corporate and environmental management plans and systems.

Box 5 — Global Warming

Although there are still many uncertainties and much disagreement among the world's scientists about the relationship between the burning of fossil fuels and the possibility of a gradual warming of the planet, it is agreed that this outcome, if it were to occur, would be serious and would irreversibly change the planet. Precautionary measures, therefore, have been put in place through international treaties and conventions in spite of disagreement among experts. The main question in recent years has been *how* precautious we should be considering the economic development benefits of burning fossil fuels in the short term.

(b) *What is the magnitude of possible impacts?* The magnitude of the environmental impact on a particular area may vary from 'negligible' to 'catastrophic'. For example, disturbance of acid sulphate soils may result in a nearby river system becoming extremely acid (pH of 1) or it may lead to a minor acidification (pH of 6). Potential impacts should be considered for both natural and human systems:

(i) Natural Systems: Impacts on genetic, species, or ecosystem abundance or diversity and processes such as atmospheric, water and nutrient cycles.

❑ negligible ❑ minimal ❑ moderate ❑ considerable ❑ catastrophic

(ii) Human Systems: Impacts on health, well-being and enjoyment of the environment, economic activities and on natural resources.

❑ negligible ❑ minimal ❑ moderate ❑ considerable ❑ catastrophic

(c) *What is the perceived value of the threatened environment?* There may be considerable difference of opinion depending on whether the affected environment is highly 'valued' or not. 'Value' may derive from importance for nature conservation, production (as in agriculture), tourism, water catchment protection, as a 'wilderness' and so on.

❑ very high ❑ high ❑ moderate ❑ low ❑ negligible

(d) *What is the temporal scale of possible impacts?* Impacts may vary in their effect in terms of time. Impacts should be considered in terms of both their *timing* and their *longevity* (or persistence):

(i) Timing: Impacts of an action/activity may be immediate or there may be a delay between action and effect. If impacts are to occur in the medium to longer term, this might be considered less serious *if* appropriate understanding of dealing with their effects can be developed. On the other hand, if uncertain irreversible impacts are delayed it may mean that large numbers of people or extensive areas of land are irreversibly affected before this impact is recognised. An example is the widespread use of a substance that turns out to be carcinogenic. A delay of up to 30 years between exposure and effect means that many people are exposed before harmful effects are conclusively demonstrated. If impacts are known not to occur until the very long term, there may be a tendency to regard them as not being the responsibility of the current generation of decision-makers. However, the principle of inter-generational equity requires us to consider the long term seriously. Because of the contradictory possibilities the effects of timing need to be assessed as:

❑ timing characteristics enhance seriousness ❑ timing characteristics neutral for seriousness ❑ timing characteristics lessen seriousness

(ii) Longevity: Impacts can persist from short- to long-term periods. It is assumed that long-term impacts are more serious than short-term impacts because the problem exists for both the environment and people of the present generation *and* future generations. An example of short-term impact (at least for the immediate human environment) is solid waste incineration compared with the disposal of nuclear waste, which has a very long-term impact. The longevity of impacts is closely related to the issue of reversibility.

❑ immediate ❑ short-term (<5yrs) ❑ mid-term (5-20yrs) ❑ long-term (20-30yrs) ❑ inter-generational

(e) *How complex or connected are the possible impacts of the action/activity?* The more connected the effects of an action/activity are to others, the greater is the chance that the environmental impacts will be more serious, irreversible or uncertain. *Complexity* and *connectivity* relate to factors such as the spatial scale and magnitude of impacts. If the environmental impacts of actions/activities are highly complex or connected, this can make it more difficult to determine the degree of certainty about their actual effects and the most appropriate precautionary measures.

(i) Complexity: The environmental impacts of some actions/activities involve very straight-forward causes and effects. Others however, have more complex interactions with the environment, with some effects triggering or feeding back into others causing multiple effects. The degree of complexity may depend on the complexity of the environment itself.

❑ discrete cause & effect ❑ complex/ multiple feedbacks & linkages

(ii) Connectivity: Some actions/activities have relatively discrete or isolated environmental impacts compared to others. The clearing of forest land for agriculture may destroy forest habitats and threaten ecological processes on a small scale. But if this clearing is conducted close to a riverbank, consideration, for example, will have to be given to the erosion of the riverbank and associated run-off of top-soil into the river which will affect water quality and the ecology locally and down-river.

❑ isolated impact ❑ several impacts ❑ many impacts

(f) *How manageable are the possible impacts?* Environmental impacts can be more or less difficult to manage. To some extent, this is related to the magnitude, complexity and reversibility of impacts. In some cases potential for impacts can be reduced to a negligible level (at least in the short term, for example, techniques for containing leachates in solid waste landfills), whereas in other cases they can be impossible to contain (for example, intractable waste). There are two aspects of the manageability of impacts to consider:

(i) Availability of means: The manageability of environmental impacts depends on whether there are already the means available to deal with impacts or whether these need to be created for this purpose. These may include alternative *products* (for example, 'fibro' made with cellulose-cement rather than asbestos-cement) or *technologies* (for example, solar/wind power rather than nuclear/ fossil fuel power). The means will vary depending on the situation from policy instruments to technologies or institutional structures and will be best developed with appropriate input from the broader community. Where there is a very serious threat, application of the precautionary principle might mean ensuring that the best available technology (BAT) is used to manage a situation. If there is a less serious threat, the best available technology not entailing excessive cost (BATNEEC) may be more appropriate (these terms are discussed more fully in Tables 1 - 6 measures).

❑ sufficient means available ❑ modified means required ❑ insufficient means available

(ii) Acceptability of means: If the means for managing environmental impacts are available, the issue is whether these means are considered acceptable. The acceptability of the means depends on recognising difficulties or objections arising from moral, political, social, cultural and economic concerns. This relates to the level and basis of public concern (see below).

❑ means acceptable ❑ modified means required ❑ means unacceptable

(g) *How much public concern is there about possible impacts?* The seriousness of environmental impact depends not only on 'objective' factors such as its magnitude, but also on the public perception about the issue. Public concern can be considered in terms of the *level* of concern and the *basis* for the concern:

(i) Level of public concern: A number of factors may affect the level of public concern. These include: the media coverage an issue receives; the perceived importance of the issue to the community (that is, health related environmental issues are generally a more 'significant' public concern than land degradation); and the perceived magnitude and spatial scale of the impacts.

❑ low ❑ moderate ❑ high

(ii) Basis for public concern: The community may have concerns about the potential impacts of an action/activity. There may be little scientific basis for this public concern, but it is important because it contributes to the overall degree of 'seriousness' in a decision-making situation. The basis for public concern may be widely shared or it may be made up of many different concerns (cultural, aesthetic, health, environmental, economic).

❑ widely shared ❑ variance in understanding ❑ many different perceptions

Q:What is the overall degree of seriousness?
(weigh up your answers to 1a - g)

❑ very serious ❑ moderately serious ❑ slightly serious ❑ not serious

(h) *How reversible are the possible impacts?* Irreversibility may seem a rather more 'black and white' concept than seriousness. But even here there is difficulty. Loss of a species is irreversible, at least at the present stage of human understanding and scientific capability. However, for many other 'threats' decision on irreversibility depends on our time-span for consideration. Many changes to ecosystems may be reversible, but only over a very long time period. Contamination of land by a toxic substance may be reversible, but again possibly only over a very long time period or involving vast costs in remediation.

Those considering precaution at the strategic or policy levels may also assess whether the implications of their decisions may be irreversible in the shorter-term. For example locking society into a particular energy source or expensive infrastructure is essentially irreversible in economic and social terms in the short and medium terms.

Q:How reversible are the possible impacts?

❑ easily/quickly reversed ❑ difficult/expensive to reverse ❑ irreversible ❑ unknown

Your Conclusion

We have assessed the *seriousness* and *reversibility* of potential threats. These now need to be combined to give an assessment of the overall 'significance' of the threat because:

- The impacts of an activity may be very serious, but may also be reversed over time (for example, collapse of fish populations through overfishing).
- An activity may be regarded as not at all serious (perhaps because we do not know much about it), but is impossible to reverse (such as the extinction of a little known insect species).
- The reversibility of action/activity may affect judgements about its seriousness — the fact that the original state of the environment cannot be easily reclaimed may suggest a threat is very serious (apparently 'little appreciated' urban bushland can become highly valued when it comes under threat of urban or industrial development).

Determining the overall significance can be complex. You must be the judge, bearing in mind that the greater the breadth of opinion, the more likely is the judgement to be widely accepted.

How Serious or Irreversible is the Threat?

Determine the overall significance of the threat (weigh up your answers for 'seriousness' (a-g) and 'irreversibility' (h)).

❑ **very significant** ❑ **moderately significant** ❑ **slightly significant** ❑ **insignificant**

If precautionary measures are needed, not only must the threat be serious or irreversible, but by definition there must be a lack of full scientific certainty about the threat. The confidence that can be placed in the evidence relating to the extent and nature of the threat is the next issue to consider.

However, even if the threat is considered insignificant it is wise to continue with this next part of Step 1: How *certain* are we about the threats to the environment? Once you assess how much confidence can be placed in scientific understanding of the threat, you may change your opinion about its potential significance.

How *Certain* Are We About Threats to the Environment?

In many cases, the environmental threats of an activity are well understood and decision-makers are clear about the outcome.

On the other hand, we may be unable to predict the exact nature of threats – the activity may be unique, or the environment it is impacting upon may be extremely complex. Often we do not know all of the components of the ecosystems we are dealing with (for example, the species present) let alone the interactions between them (that is, ecological processes). There may be:

- uncertainty about particular aspects of the threat;
- aspects of the threat that we are unable to predict because we do not have the tools available or the level of knowledge required (that is, these cannot be determined).
- threats that we are not even aware could exist (that is, we may be ignorant).

In these cases, decisions are more difficult to make, but the precautionary principle requires us to try to avoid damage that *might* be serious or irreversible by putting in place *precautionary measures*.

Precaution or Prevention?

Remember, the precautionary principle is about those situations in which there is a lack of full scientific certainty. So when we are *confident* about the threats, it is *preventative* measures that are required. In these cases we must weigh up (i) the assessment of risks and costs of environmental damage versus benefits; and (ii) costs of *prevention* versus benefits.

However, where there is scientific uncertainty

regarding the threats we are required by the precautionary principle to consider *precautionary* measures. The extent of precaution that is considered appropriate is likewise influenced by the various costs and benefits involved, although these are likely to be more difficult to estimate given the inherent uncertainty.

How to Decide About Uncertainty?

When assessing how certain you are about a threat, the following actions and considerations may be useful.

1 *Involve a range of views in assessing the degree of certainty about threats to the environment*

The involvement of people from a variety of relevant disciplines and the broader community will enhance the chances of identifying uncertainties. This is particularly necessary in situations involving complex ecological interactions. It will also help to facilitate 'consensus' when making your assessment of the degree of certainty about environmental threats. The use of structured discussion sessions, such as 'planning focus' or 'value management', aimed at identifying the range of possible factors influencing a situation and exploring the possible solutions, is likely to be useful.

Use the lists of points and questions in 2 and 3 to prompt discussion when involving others. These should help to draw out any differences in opinion and to bring out issues you may have forgotten or not considered.

2 *Assess what you do know and what you don't know about the threat*

Determine how well you understand the nature of the threat by considering:

- how the activity is likely to impact on the environment;
- the full range of possible impacts;
- whether there is a discrete impact or a complex web of effects;
- whether there are cumulative effects;
- whether there are synergistic effects;
- the magnitude of impact;
- the spatial scale of impact/s;
- the time during which impacts may occur – immediate or delayed;
- whether effects are short-term or persistent;
- the nature of the environment that may be affected by the activity;
- whether different ecosystems will react in the same way;
- what factors may influence interaction between the environment and the activity?

Use the above as a checklist.

3 *Assess the methods used to understand the threat*

Assess the reliability and appropriateness of the methods used to understand the threat. Use these points as a checklist:

- have we made the right assumptions about what is relevant for our analysis?
- have we used the right models to describe possible threats?
- is the experimental design appropriate?
- are data collection methods reliable?
- is our process of data analysis suitable to answer our questions?
- can firm conclusions be drawn from the analysis?
- is there consensus on these conclusions?

By now you have identified and considered many aspects of the possible threats associated with your activities. You should have decided:

- how well you understand these threats
- how you have 'defined' these threats
- whether all relevant factors have been considered.

With this information in mind, now assess the *degree* of uncertainty.

4 *Assess how certain you are about the threat*

(a) *Is there sufficient evidence to justify precautionary action?* Can we be 'reasonably confident' that the available evidence suggests there could be serious or irreversible environmental harm? It would probably be 'unreasonable' to apply precautionary measures on the basis of mere speculation. But we do need to

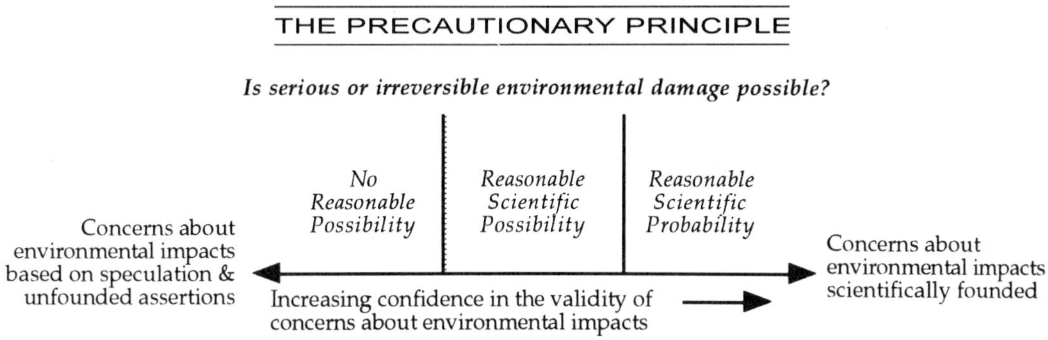

Figure 2: Levels of confidence in evidence about possible threats (adapted from Hickey and Walker, 1995)

keep an open mind since there are now many examples of 'surprise' environmental impacts. The hole in the ozone layer as a result of the action of CFCs is one such example. We need to take account of what may be our present ignorance.

Three levels of confidence can be used to judge whether it is scientifically 'reasonable' that there *might be* serious or irreversible environmental impacts associated with a particular activity (see Figure 2). Even though there may still be uncertainties about the details of the impacts, this will help you decide whether concerns about possible impacts are scientifically based or not.

No reasonable possibility exists when there is no scientific evidence that would support a hypothesis about environmental impact, or when there are strong reservations about the accuracy of the evidence or the methods and assumptions used to collect and interpret the evidence. The hypothesis is considered scientifically *un*founded.

Reasonable scientific possibility exists when scientific evidence (as opposed to mere hypotheses, speculation or intuition) provides a rational basis that warrants drawing the conclusions from the evidence, even though reasonable scientific experts may disagree on whether that conclusion is the only valid one based on the evidence. The hypothesis is scientifically plausible.

Reasonable scientific probability (or likelihood) exists when scientific experts generally agree that the available evidence and methods used to interpret it are valid and reliable and when there is also 'general acceptance' by the relevant scientific community of the specific conclusions drawn from the evidence. The hypothesis is scientifically likely.

Q: Is it scientifically 'reasonable' that there might be serious or irreversible environmental impacts?

❑ no reasonable possibility ❑ reasonable scientific possibility ❑ reasonable scientific probability

If it is thought that there is a *reasonable scientific possibility* or *probability* that there could be a significant threat, you now have to consider the *level* of uncertainty.

If you think there is no reasonable scientific possibility or probability, remember there is always *some* uncertainty about threats to the environment because of the complexity of ecosystems. So there is *always* a need for *some* precautionary measures when it is suspected that threats *might* be serious or irreversible. Hence, even if it is thought that there is *no reasonable possibility* that there could be a significant threat, give the environment the benefit of the doubt and consider if there is ignorance about the nature of the threat, or aspects which cannot currently be determined. In other words, consider other *types* of uncertainty. See Box 6.

There are different 'shades' of certainty and knowing how confident we can be in predicting impacts helps to guide us in making decisions about our actions — it indicates how precautious we should be.

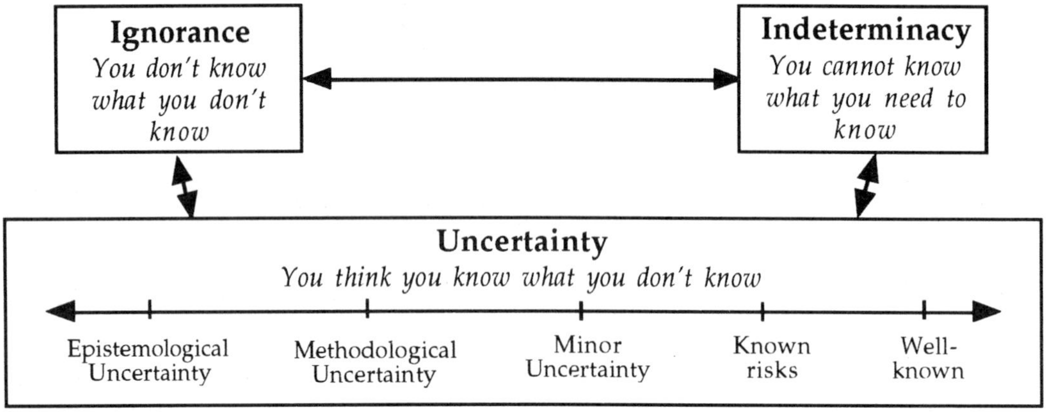

Figure 3: Levels and Types of Uncertainty

(b) *What is the level of uncertainty?* If we have decided that there is sufficient evidence of a significant threat and hence that precautionary measures are required, we need to decide *how much* precaution is appropriate. This will be influenced by the level of uncertainty (in combination with the significance of the threat). Generally, the greater the level of uncertainty, the greater the precautionary measures we should apply.

So our next step is to assess the level of uncertainty. We can plot this on a scale of understanding which extends from 'well-known' through to 'epistemological uncertainty' (See Figure 3 and definitions below).

However, uncertainty is a complex concept. It cannot always be ranked from low to high. In some situations we simply 'don't know what we don't know'. We are ignorant! Or we don't know what questions to ask. These types of uncertainty cannot be ranked on a scale; rather we need to recognise that there is always a context of possible ignorance and indeterminacies surrounding our level of certainty about our activities (see Figure 3 and Box 6).

Well-known: If the impact of an action/activity is well-known, there is an overwhelming consensus about the outcome and decisions are made on the basis of a high degree of certainty. The precautionary principle is not relevant if outcomes are well-known (*but* see Box 6 below). Rather we must decide whether to put in place *preventative* measures.

Known Risks: If a range of likely outcomes and their probability of occurrence can be predicted with confidence (using structured risk assessment techniques), we are then concerned with a set of known *risks* rather than with 'uncertainty'. In this case *prevention* rather than precaution needs to be considered (*but* see Box 6).

Uncertainty: If only the general nature and direction of likely impacts are understood, that is, the parameters of the situation but not the odds, decisions are made in the face of uncertainty. It is at this point that the precautionary principle begins to apply. We can distinguish three kinds of uncertainty:

Technical uncertainty concerns the accuracy of the data (precision of observations and measurements) or the conclusions drawn from the data. The appropriateness of the methodology used is not in question.

Methodological uncertainty concerns assumptions behind the scientific methods used to collect and interpret the data and whether the data set is appropriate or the right analytical tools are being applied.

Epistemological uncertainty concerns whether the knowledge framework for the resolution of problems is correct.

Box 6 —Ignorance and Indeterminacy

Ignorance

You think there is *no reasonable possibility* of a significant threat, there are two possibilities to consider:

1. That there might be *no* threat and the precautionary principle is not relevant to your decision-making.
2. That we might be ignorant and therefore we:
   - fail to recognise the existence of a threat;
   - fail to recognise all aspects of a threat; or
   - that we do not currently have the means to understand the nature, the cause, or the possibility of such a threat (that is, indeterminacy, see below).

In other words, there may be different *types* of uncertainty that make it currently impossible to recognise that there actually is a threat of serious or irreversible environmental damage.

Indeterminacy

If we cannot determine the accuracy of the scientific and social assumptions on which our assessment of a threat is based, this is referred to as 'indeterminacy' .

Science can define risks or uncertainties by 'freezing' a surrounding context which may or may not be this way in real life. The resulting knowledge is therefore conditional knowledge depending on whether these assumptions turn out to be valid. But this is indeterminate since we don't know all the factors influencing the causal chains or networks. That is, these networks or chains are open rather than bounded or closed as in cases of technical uncertainty.

Indeterminacies can arise from:

- not knowing whether the type of scientific knowledge and the questions posed by scientists are appropriate and sufficient for the circumstances;
- the social context in which that knowledge is applied through policy commitments;
- the stability of that social context (from Wynne, 1992).

Consider a situation where an assessment of threat includes environmental protection through the reliable operation of an environmental management system. If the social context changes, such that the EMS is no longer reliable, the assumptions on which we have based assessments of the threat have changed. The threat may now be much greater. The operation of nuclear facilities in some Russian states following the collapse of the USSR is one example.

Both ignorance and indeterminacies may exist at all times in our knowledge about the effects of our activities, no matter how certain we think we are, because we cannot know what we do not know. Suggestions or theories that may seem scientifically unfounded now, may well turn out to be quite valid in years to come. *There is always a context of possible ignorance and indeterminacies surrounding our level of certainty about our activities* (see Figure 3).

This *does not mean* that development should halt because we *cannot* know what the effect will be. But it *does mean* that we should try to be precautious at all times, even when we think we know how certain we are about the impacts of our activities. Remember, the precautionary principle requires us to give the environment the benefit of the doubt.

Ignorance and indeterminacy should not stop us from acting for the fear of threats that we cannot foresee, because after all, there may turn out to be no threat. But some types of precautionary measures can and should still be taken (see Step 2).

Q: What is the level of uncertainty?

The *level* of uncertainty will help you decide how much precaution is required (discussed in Step 3).

❑ epistemological uncertainty ❑ methodological uncertainty ❑ technical uncertainty ❑ known risks ❑ well-known

Even if you have considered the level of uncertainty to be low, consider that other *types* of uncertainty may be possible and other types of precautionary measures may still be required. Go to Box 6.

(c) *Consider the potential to reduce uncertainty.* One response to the precautionary principle is to attempt to reduce uncertainty so that the application of precautionary measures is unnecessary. Indeed it may be argued that reducing uncertainty is in itself a precautionary measure!

If scientific uncertainties about the potential threats of activities can be reduced to a limited range of likely outcomes (known risks) then decision-making about these activities involves risk management and the need for precaution is reduced. If it is scientifically feasible and socially and economically acceptable to reduce uncertainties to this level, then it is desirable to do so. Consider how feasible it is to reduce uncertainties to the level of 'risk' rather than uncertainty:

In principle — The capability of current scientific practice will determine what is possible 'in principle'

Economically — The availability of finance to reduce uncertainty must be considered. Ideally, the precautionary principle places the burden on the proponent to fund the reduction of uncertainty through scientific studies and other means.

Within a 'reasonable' time-frame — What is a 'reasonable' time frame depends on many variables that a range of people (including proponents) must determine. For example, reducing uncertainty in understanding of ecosystem function may require studies over many years in order to appreciate the existence of natural variability. Developments are not likely to be feasible if this is required.

Q: What is the overall feasibility of reducing uncertainties about impacts to the level of a set of known risks?

❑ unfeasible ❑ feasible ❑ highly feasible

If it is feasible to reduce uncertainty (that is, economically and within a 'reasonable' time frame), you must decide whether to:

- attempt to reduce uncertainty and, if successful, introduce a program of risk management or,
- implement precautionary measures appropriate to the present level of uncertainty (see Steps 2-4).

If it is not feasible at the present to reduce uncertainties to an acceptable level, precautionary measures should be applied (see Steps 2-4).

How Certain Are We About Threats to the Environment?

Taking account of both the level of uncertainty and potential for its reduction, make your final assessment of the level of uncertainty.

❑ epistemological uncertainty ❑ methodological uncertainty ❑ technical uncertainty ❑ known risks ❑ well-known

Review the Situation

If you assess the situation incorrectly in the first place and then take action on that basis, what are the likely environmental/ecological consequences? legal consequences? financial or economic consequences? public image or public relations consequences?

It is now important that you review your answers to the two questions: Are threats serious or irreversible; and how certain are we about the threats to the environment by considering:

Basis for assessment

- On what information are your answers based?
- Are these widely accepted views?

— by the scientific community?
— by the general public?

Public participation and understanding

- Have you deliberately involved outside stakeholders in your assessment?
- Have you taken steps to empower them to participate meaningfully?

Remember, others may bring valuable perspectives to the assessment that you have not considered.

Are Precautionary Measures Needed?

1. Is there a threat of significant (serious or irreversible) environmental damage?

   ❑ **Yes** ❑ **No** ❑ **Maybe**

2. Is there scientific uncertainty ?

   Remember that for precaution rather than prevention to be appropriate, the uncertainty should be greater than a set of known risks

   ❑ **Yes** ❑ **No** ❑ **Maybe**

YES

If the answer to both questions is *yes*, you need to use the precautionary principle to guide your decision. The remaining Steps discuss how to apply the principle.

NO

If the answer to one or both questions is *no*, consider whether you have adequately assessed the situation. If you consider application of precaution is not relevant in this situation, can you justify why this is so? Make sure you are assessing the possibility of threats to the environment fairly, that is, giving the benefit of the doubt to the environment rather than to the economic bottom line. Applying the precautionary principle does not necessarily mean lost income. In fact, being precautious with the environment can save money in the long term because of the possible costs of compensation, legal fees and environmental repair. Remember, under the IGAE and a range of Federal, State and local laws, you are obliged to apply the precautionary principle if it is appropriate to do so.

MAYBE

If the answer is *maybe*, you should either assess the situation again, or, go on to the remaining Steps and use the precautionary principle to guide your final decision anyway. ***It is better to be safe than sorry.***

Step 2: How Precautious Should We Be?

Steps 2 and 3 should be considered together. The amount of precaution tht can be applied will depend partly on the precautionary measures available, while at the same time the precautionary measures applied will be determined by how much precaution is needed.

How precautious we should be depends on how 'significant' (that is, serious or irreversible) the threat is and how much uncertainty there is about the threat.

There are two 'rules of thumb':

1. The more *significant* the threat, the greater the degree of precaution required.
2. The more *uncertain* the threat, the greater the degree of precaution required.

Figure 4 gives you a guide. Plotting the *significance* of the threat (vertical axis) against the degree of *uncertainty* about the threat (horizontal axis) gives an *indication* of how much

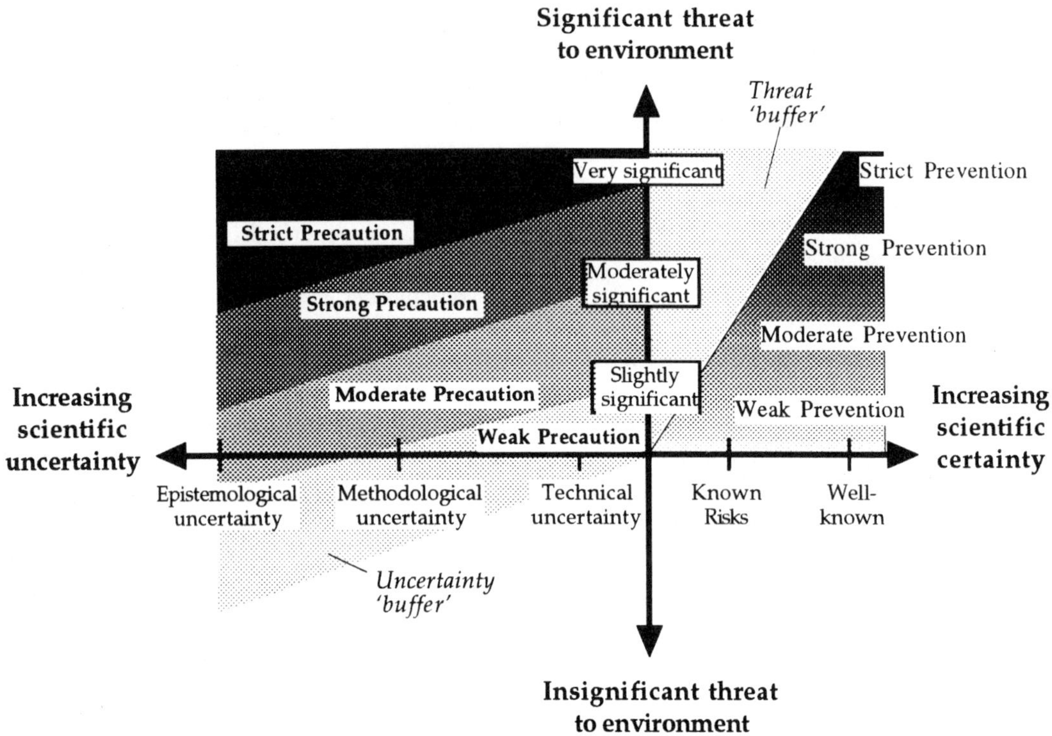

Figure 4: Determining Degrees of Precaution

precaution is appropriate. For example, if there is only minor or technical scientific uncertainty, strict precaution may not be appropriate unless there is also a very significant threat to the environment. But if the degree of uncertainty was much larger, then strict precaution might be appropriate, even if the threat was only considered to be moderately significant. Major uncertainty may arise either because of large gaps in knowledge about key aspects of the environment, or because we are uncertain whether the methods used to investigate the potential threats are appropriate.

The extent of the threat is described on the vertical axis. The intersection with the horizontal axis marks the threshold for 'serious or irreversible' environmental damage — which defines the need for precaution.

Scientific uncertainty is described on the horizontal axis. The intersection with the vertical axis marks the threshold for sufficient scientific uncertainty to warrant precaution.

Remember, that indeterminacies and ignorance are other *types* of scientific uncertainty that need to be considered. These are not shown on the horizontal axis because they can occur right through the 'spectrum of uncertainty'. Rather, the entire graph is placed within the context of 'ignorance and indeterminacies'.

The top-left quadrant is the area defined as requiring precautionary measures. But because of ignorance and indeterminacy 'precautionary buffer zones' have been added. The 'threat' buffer in the top-right quadrant accounts for the danger of 'getting it wrong' when the *threat* is very large. The 'uncertainty' buffer in the bottom-left quadrant accounts for the danger of 'getting it wrong' when *uncertainty* is very large. With increasing uncertainty, judgements made about the likely threat may need to be revised.

The top-right quadrant describes a situation where preventative measures are in order — threats are significant and there is considerable confidence that our assessment is correct. The 'threat' buffer should encourage a precautionary approach to decisions on prevention in this zone. The bottom-right corner of the graph describes a situation where uncertainty is low and any environmental change associated with an activity is thought to have a very low threat of damage. Here we may simply use cost-benefit analysis to determine our best course of action. The bottom-left quadrant describes situations where uncertainty is high but threats are not deemed significant. It may cover situations where activities in highly modified ecosystems are aimed at environmental restoration.

Use Figure 4 to help decide how precautious you should be in relation to the threat assessed in Step 1.

Take care in assessing how much precaution is appropriate. There are obvious dangers in underestimating the amount of precaution needed, particularly if the situation involves a high degree of irreversibility. In general, a 'rule of thumb' is that *too much precaution is better than too little,* because we can always gradually reduce precaution as more scientific evidence about the harmfulness of activities or new technologies or practices becomes available. A careful, cautious approach to your assessments is very important.

How Precautious Should We Be?

Look back to the answers you gave in Step 1. Try to 'plot' significance and uncertainty to determine the appropriate degree of precaution, using Figure 4 as a guide. Move up the vertical axis to your answer to the *significance of the threat.* Then move across horizontally until you reach the appropriate *degree of certainty* about this threat.

❑ **strict** ❑ **strong** ❑ **moderate** ❑ **weak** ❑ **precautionary buffer**

Step 3: What Precautionary Measures *Can* You Apply?

Precautionary measures can be direct and indirect. It depends on the situation.

Direct Precautionary Measures

Measures can be implemented to avoid threats of serious or irreversible environmental damage that might arise as a result of *specific* actions, activities or projects that are already occurring (such as operating a landfill, harvesting fish or timber or constructing a mine), or activities *proposed* for the future (such as the development of a suburb or of new technologies, products or processes). While different responses are appropriate in different situations, some generic examples of directly precautionary measures are:

- to stop the activity or modify it to significantly lessen possible environmental impacts;
- to reduce the uncertainties so potential impacts can be re-evaluated,
- to use rigorous quality systems to limit impacts during operations (for example, to ensure responsible practices are conducted by contractors working in sensitive forestry areas, or that discharges into waterways are conscientiously conducted by plant operators, well within regulations);
- to ensure that any environmental change that does occur is monitored closely and the activity is halted before the damage becomes serious or irreversible.

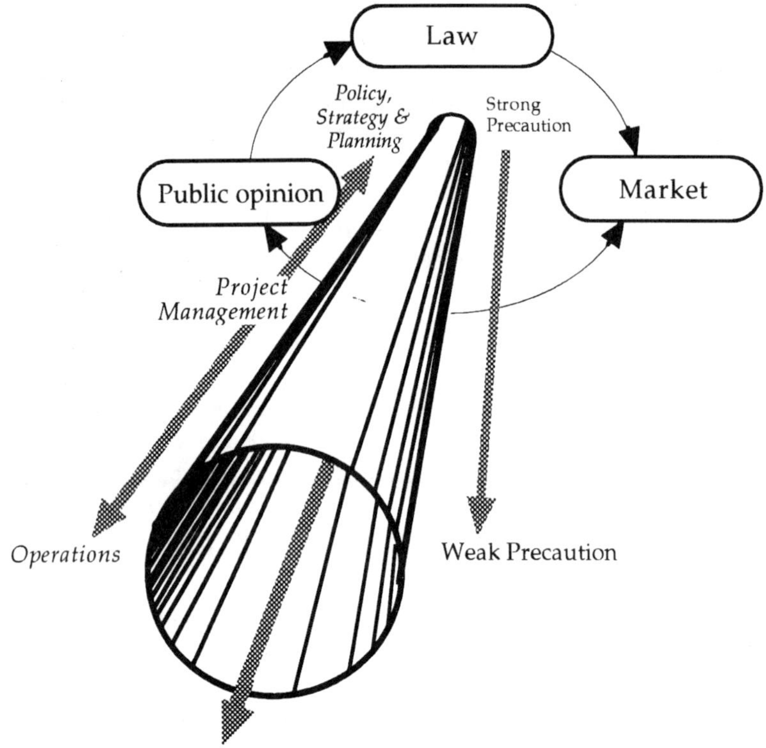

Figure 5: The Decision-Making Pipe

Indirect Precautionary Measures

But measures can also be applied to more generally *facilitate* the implementation of the precautionary principle. Figure 5 illustrates the flow of decision-making as a 'pipe' running from strategy and policy down to operations.

For any decision we have to make about a particular project or activity (such as whether to renew a particular fishing licence), there have been other more general decisions made further up the decision-making 'pipe' (for example, about laws and policies). These higher level decisions limit the precautious actions that are possible for decision-makers down at the operations end of the pipe. For example, timetabling for a logging operation set by project managers may create pressures on contractors, encouraging less careful operations to meet deadlines. Further up the pipe, decisions about allocating different types of land-use may have been taken which has meant that in particular areas selected for logging, the likelihood of serious or irreversible environmental damage is particularly high. Again, this creates pressures on those working at the operations end of the pipe.

Direct precautionary measures can be applied to particular actions or activities to avoid threats of potentially serious or irreversible environmental damage. Indirect precautionary measures can be applied to facilitate the avoidance of threats in the longer term by encouraging behaviour that produces precautionary results.

We need to look further up the decision-making 'pipe' and introduce a range of indirect precautionary measures wherever possible. This should help to reduce the number of specific situations that require precautionary measures (such as the closing down of a particular fishing area) by preventing threatening situations from arising in the first place. In a sense, these more strategic measures are the most precautionary of all. So, a policy that helps to minimise the amount of solid waste generated in the community is more precautionary than a policy of reducing impacts at the point of disposal through landfill or incineration, because it reduces the overall level of potential environmental threat in the first place.

Some generic examples of indirect precautionary measures might be to:

- develop organisational policies and management approaches that systematically consider long-term environmental impacts before proceeding with activities, or
- implement policies, laws and other incentives and that will effectively reduce amounts of waste materials and energy currently generated.

Although decisions taken further up the pipe are generally more significant for the application of the precautionary principle, that does *not* mean that responsibility for application of the principle should be merely passed up the pipe. If solid waste disposal methods like landfill and incineration are necessary then these operations must be conducted with precaution through containment techniques and emissions controls, and further up the pipe, by separating recyclables and non-recyclables at their source. Responsibility for application of the precautionary principle should be shared and may apply at many levels simultaneously.

Figure 5 also shows the relationship between the legal system, the market and public opinion. How does this affect decision-making? Through the process of elections and political lobbying, public opinion is gradually being reflected in the legal system while the market place is partly also a reflection of public opinion (that is, through 'consumer choice'). The legal system through regulations can have direct influence on the market. These three 'spheres' together can directly and indirectly influence public and private decision-making. These influences are increasingly shifting the focus of environmental regulation away from the operations end of the pipe towards a generally more precautionary environmental management approach.

Responsibility for the precautionary principle should be taken at all levels by adopting an integrated approach to environmental management within private-sector organisations and all levels of government administration.

The increasing inclusion of the precautionary principle into legislation and policy is making this a requirement for organisations.

Which Measures?

The types of measures you might apply depend on the sorts of decisions you are responsible for making (see Figure 6). If you are concerned with decisions about (1) existing operations and ongoing day-to-day activities or with (2) the development, assessment or the approval of a proposed project or activity, you will be applying *direct precautionary measures* to avoid specific threats to the environment. But if you are involved with decisions concerning (3) organisational management, (4) administrative procedures, (5) legislation and law reform, or (6) broad policy, strategy and planning, a range of *indirect precautionary measures* will be relevant. The sorts of measures that can be applied at these six levels of decision-making are shown in Tables 1-6 in Part 3.

Measures to Deal with Ignorance and Indeterminacy

Indeterminacy and ignorance cannot be excluded from environmental decision-making. But, how can we possibly be precautionary about threats

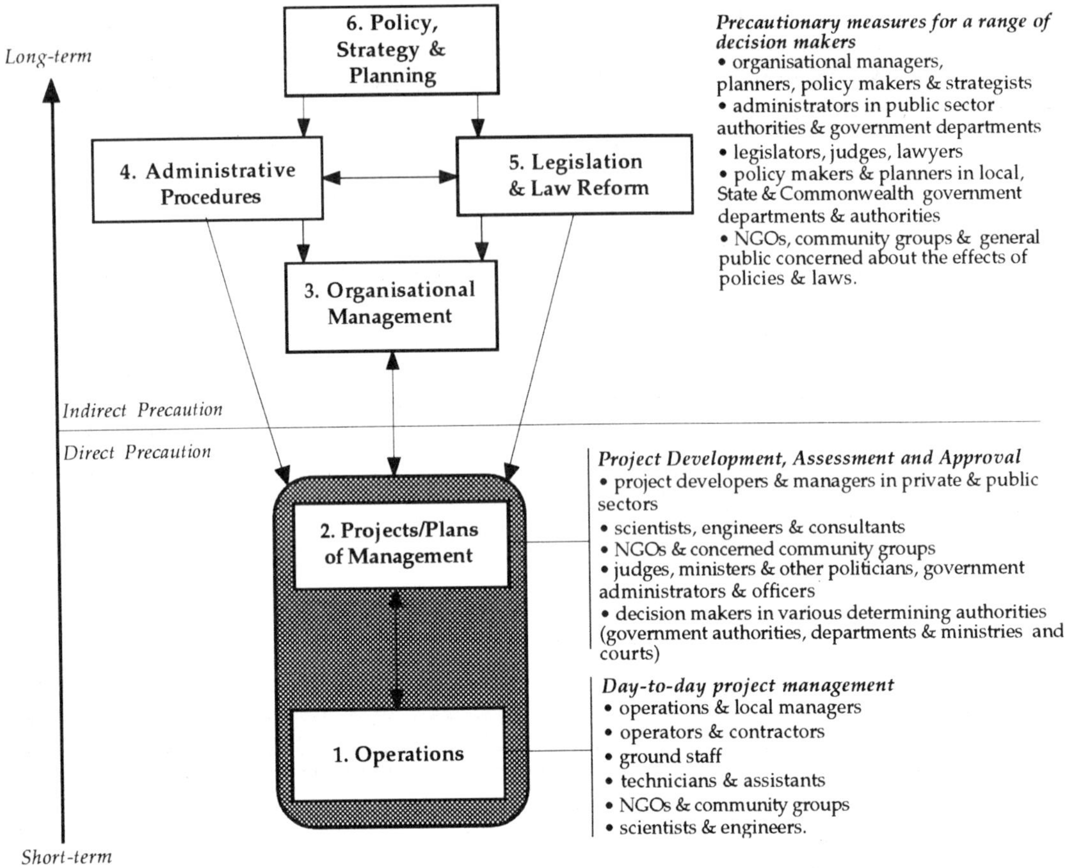

Figure 6: Different precautionary measures and levels of decision-making

we do not have the means to understand or that we don't even know exist?

There are some general measures that may assist in reducing the potential for serious or irreversible environmental damage arising from ignorance or indeterminacies. These are:

- Recognise the importance of basic scientific research across all fields of science. This will help increase our understanding of the environment and the nature of our interactions with it.
- Emphasise the importance of monitoring. Increased research, however useful, will not provide full understanding. Monitoring offers the chance for early recognition of serious or irreversible impacts and possibly the opportunity to take action to deal with these.
- Be open minded! History can provide many examples of cases where 'outlandish' propositions have been proven true.
- Promote and encourage attempts at all levels to shift from our present unsustainable framework for development to one which is ecologically sustainable. This will require a considerable shift in ethos.

Consider which measures you can apply by looking at Tables 1-6 in Part 3.

Step 4: What Precautionary Measures *Should* Be Applied?

In Step 2 we considered *in principle* how much precaution is appropriate. We now need to make a final decision on precautionary measures by setting this 'in principle' decision within a broader context.

Developing Options

We need to develop a range of options based on:

- What precautionary measures are available (Step 3) and their social and economic costs;
- How much precaution is appropriate (Step 2);
- Consideration of the other principles of ESD (refer back to Box 1). ESD principles are intended to be applied together as a 'package'. This means that precautionary measures must be 'appropriate' in the context of other ESD principles including: inter-generational and intra-generational equity and the conservation of biological diversity.

Of course, with this range of considerations, there is no one set of measures that is 'correct' and different options will satisfy these criteria to varying degrees. The Intergovernmental Agreement on the Environment requires that we decide on precautionary measures by *assessing the risk-weighted consequences of various options.*

Assessing Options

In assessing the risk-weighted consequences of various options, try to make a judgement about how much precaution is appropriate knowing *there are dangers in applying too little precaution, but also in applying too much precaution (see Box 7).*

Choosing an option

Having assessed the set of risks associated with each option, a decision must be made.

Remember! Judging the appropriate amount of precaution and developing the appropriate precautionary measures are best achieved with the input of a broad range of people.

If the process for developing and selecting options is not 'transparent' or does not involve the full range of people with an interest in the outcome, the decision is much more likely to be regarded with mistrust and possibly even community resistance.

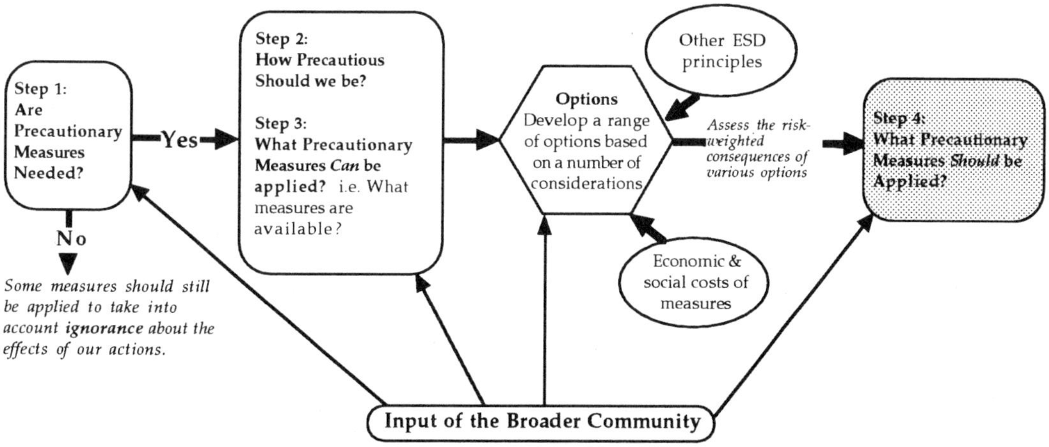

Figure 7: Choosing appropriate precautionary measures

Box 7— Too Much Precaution

Too Much Precaution?

'There are some dangers with getting too carried away with the application of precaution at any cost. In the absence of comparative risk assessment, the consequences of curtailing potentially beneficial activity and creating another set of unforeseeable risks for an unprepared society could be greater than proceeding step by step with prudent precaution.' *(O'Riordan, 1994)*.

A Need For Strict Precaution?

Strict precaution applies when there is good reason to suspect there is a threat of very serious or highly irreversible environmental damage, and/or where there are significant scientific uncertainties about this damage. *In general, strict precaution is about ensuring the reversibility of decisions*. In *most* cases the appropriate precautionary response would be to stop the activity from continuing or to suspend its approval until enough information has been gathered to reduce uncertainties. The degree of strictness may be reduced and prohibited activities might be allowed at a later stage if the state of knowledge improves enough to indicate that serious or irreversible damage is unlikely. But if weak measures are applied *before* strict measures, there may be no opportunities to reverse any damage that is done.

Does Strict Precaution Always Mean Saying No?

Of course not. It is not always possible to simply stop or prevent an activity on the grounds that it *may* cause significant environmental damage. There may be very strong economic and social reasons for trying to find a way to allow the activity to proceed. For example, a strictly precautionary option may be to only allow continuation or approval of an activity if it is conducted in stages, within very tightly defined boundaries that should allow full reversibility.

If an assessment of a situation suggests that strict precaution is appropriate, follow these 'rules' or tips for allowing the activity to proceed:

- There are appropriate justifications for the activity, considering benefits beyond just short-term economic ones. This justification should be sought by open negotiation, with provisions for freedom of information, broad representation and independent data reliability statements concerning predicted impacts where appropriate.
- All options and feasible alternatives for achieving the same goal have been considered carefully.
- There are very high social costs if the activity does not go ahead and/or society is prepared to accept the predicted level of possible damage (recognising also that damage may be higher than predicted if uncertainty is very high).
- Big ecological safety margins have been included, so that first indications of any possible damage will be visible *before* they become serious or irreversible and further precautionary actions can be taken accordingly; and,
- There is significant insurance or enforceable compensation mechanisms (bonds) in place if other precautionary measures should fail or impacts are considerably more significant than predicted.

CASE STUDY
Ground Water Management: the Tomago Special Area

Mining and groundwater

The Tomago Sandbeds, near Newcastle in New South Wales, provide groundwater which can contribute up to 25 per cent of the Hunter Water Corporation's annual supply of water to the Newcastle district. The Tomago Special Area, is gazetted for protection of the groundwater resource for this use.

Since the early 1970s, Rutile and Zircon Mines Ltd (RZM) has been extracting minerals from these sandbeds under a mining lease from the Department of Mineral Resources. The mining activity is known to impact on groundwater quality mainly by increasing iron and arsenic levels. This impact increased after 1990 when the company mined at a level which extended deeper into the aquifer.

Groundwater is extracted as part of RZM's mining process. The extraction of groundwater was regulated by the Hunter Water Corporation until 1995 when this role passed to the Department of Land and Water Conservation (DLAWC). Because of the essential role of groundwater extraction in the mining process, continuation of mining depended on the DLAWC issuing a permit for groundwater extraction. DLAWC's decision to issue such a permit took place in the context of public concern over degradation of groundwater resources in the region and more generally the need to apply the precautionary principle in making the decision.

The issuing of the permit is used here as a case study in the application of the precautionary principle by a government agency. The decision is examined in a broader policy context with reference to the levels of decision-making used throughout this book and which provide the headings for Tables 1 to 6. See Figure 6.

Policy, Strategy and Planning & Administrative Procedures

Principles for groundwater management in New South Wales are established in the Draft State Groundwater Policy (Framework Document). This gives commitment to (1) sustain the long-term beneficial uses of the State's groundwater resources and (2) ensure the needs of environmental systems dependent on groundwater are met. It also sets out the *process* for implementation of the policy at local level, requiring the development of Groundwater Management Plans where groundwater systems are stressed or contentious. These plans integrate management of groundwater through co-operation of State Government agencies, local government and the community.

The emphasis on sustainability in this policy sets the scene for a precautionary approach at the highest level (level 6 in Figure 6). The policy also sets out administrative arrangements (level 4 in Figure 6) which can be used to reinforce a precautionary approach by ensuring all areas of potential concern are covered in an integrated manner across government and with community involvement.

Projects/Plans of Management

The *Tomago-Tomaree-Stockton Groundwater Management Plan* is one of the first of such local plans to be developed under the State Groundwater Policy. DLAWC has described the Plan as a *living document* which will be regularly reviewed to take account of better knowledge about the groundwater system and changing community attitudes. The Plan includes:

- a management objective to ensure sustainability of both quality and quantity of groundwater resources for present and future generations;
- establishment of a management objective to ensure that the needs of environmental systems dependent on groundwater are met;
- a proposal that all potential sources of contamination monitor water quality at their site boundaries;
- requiring RZM to monitor impacts of current heavy mineral extraction practices, and to modify activities which degrade or threaten the long-term groundwater quality;
- a proposal that the NPWS identifies wetlands and other sensitive vegetation that depend on and/or protect groundwater quality and quantity.

RZM's operations are just one of a number of activities in the area covered by the Plan.

Of more specific relevance for this case study is the *Regulation 27(1) Permit* through which DLAWC regulates the operations of RZM with respect to groundwater extraction. In keeping with the State Groundwater Policy, public concern over loss of groundwater resources in the area and DLAWC's internal promotion of the inclusion of the precautionary principle in decision-making and legislative review, the Department aimed to adopt a precautionary approach in determining the issuance of the permit.

Given the knowledge that mining was increasing iron and arsenic levels in the groundwater and that the long-term effects of this were uncertain but potentially significant, a strongly precautionary approach would have required cessation of the groundwater extraction (and hence mining) until these effects were able to be better determined.

However, the precautionary principle as defined in the IGAE has conditional clauses which take account of the broader context. Bearing in mind the on-going nature of the operations through 20 years of approval by Hunter Water for groundwater extraction, DLAWC issued a groundwater extraction permit to RZM in August 1995 to enable mining to continue uninterrupted for the present, *but* certain precautionary measures were built into the permit as follows:

- The permit was issued for up to 5 years with a *'floating term' condition* allowing for mining to be stopped by the Department on 7 days notice if the Department believes this is necessary or desirable for immediate protection of the groundwater.
- The Company was required to investigate the effects of mining in order to resolve the existing uncertainty. More specifically, within the first year of the Permit, RZM was required to prepare a *Management Strategy* reviewing the impacts of mining operations on any aquifer or water resources in the area. The Strategy had to outline a methodology and rationale demonstrating how mining *did not/ would not* cause the quality of the groundwater to deteriorate below acceptable water quality limits in the medium to long term.

  In other words the onus was put on the Company to demonstrate 'acceptable impact' rather than others having to demonstrate unacceptable impact.

- A *Steering Committee* approved Terms of Reference for the Strategy, oversaw its development and ensured that the investigation program of the Strategy was consistent with the objectives of the Permit. The research methodology and interpretation of research findings were closely scrutinised by the Steering Committee.

  The Steering Committee was made up of representatives of the relevant government agencies and RZM but lacked community representatives. Although its work was primarily technical, inclusion of representatives with appropriate technical background chosen by environment or community groups, would have made for a potentially broader spectrum of views. This would have been more in keeping with the precautionary principle.

- Vegetation and habitat were required to be managed in accordance with a *Procedures Manual.* The Manual was to be ratified by a Committee comprising members of relevant government agencies.
- RZM was required to use *Best Practice* in its mining operations.
- RZM had to undertake *restoration/ rehabilitation* on the site after operations ceased.

Referring to Figure 6, these last three requirements although emerging from Projects/ Plan of Management (level 2) are implemented through Organisational Management (level 3) and Operations (level 1).

Outcomes

The study findings were released on 22 February 1996. On the basis of the evidence presented, DLAWC concluded that:

- there was not good reason to withdraw the Permit issued to RZM;
- the mining activities are intensifying a natural process of movement of metals into the groundwater and this will return to natural levels within 20 years of cessation of the mining (likely to end in 1999);
- the effects are likely to be restricted to the area within the mining leases (16 per cent of the whole aquifer area) which are valid to 2015;
- the time frame from likely cessation of mining to expiry of the lease allows time to assess whether site rehabilitation and water quality require further restoration work by RZM. If so, significant security bonds required of RZM can be used.
- while the mining will alienate a portion of the groundwater in the short to medium term, neither the Hunter Water Corporation nor other water users are expected to be disadvantaged in the long-term;
- RZM's operating practices have been required to change in a way that is believed to significantly reduce degradation of the groundwater compared to that occurring in the past. This has been at considerable cost to the company.
- although the activities of RZM are regulated by 3 government agencies (EPA, Department of Mineral Resources, DLAWC), these agencies have developed an integrated system for regulating the mining process. *A precautionary approach is evident in the requirements of these three agencies.*

Conclusion

This case study demonstrates a pragmatic approach to application of the precautionary principle in relation to protection of groundwater in the Tomago Special Area in New South Wales. The application of precaution is integrated over a number of levels of decision-making as shown in Figure 6. These include strategic policy, administrative procedures, project approval and plans of management, intra-organisational management and operations.

Precaution could have been applied more strongly, and indeed public concern in this regard has been loudly voiced. However, it is important to note that mining already existed without a groundwater management plan. Hence, the decision to allow mining to continue has been set within the broader context of social considerations which form the second part of the IGAE definition of the principle.

Precaution has been built into the decision by moving the onus onto RZM to demonstrate that long-term harm is unlikely, and by stipulating conditions relating to operational management and to the monitoring of operations so as to give an early warning of any harmful effects. Finally in case rehabilitation is required beyond any that is expected, a significant bond has been required from RZM.

PART THREE

PRECAUTIONARY MEASURES

Part 3 provides suggestions for a broad range of precautionary 'measures' that could be considered. It is organised into six Tables corresponding to the six levels of 'decision-making' shown in Figures 1 and 6. These range from the 'operations' level in Table 1 through to 'Policy, strategy and planning' in Table 6.

Remember from the discussion so far that the scope for precautionary action becomes greater moving up the decision-making 'ladder'. Decisions at higher levels (for example, policy formulation, planning) set the framework and requirements for decisions and actions at lower levels (for example, operations). However, this does not absolve those at the operations level from a role. Indeed the reverse is true since at this level decisions taken within the framework set at higher levels may be more or less precautionary. As well, astute observation provides a vital feedback to decision-making at higher levels. Consider the recent, widely acclaimed, rescue of stranded yachtsmen in the Southern Ocean off Australia. Although procedures were set at higher decision levels, the success of the operation ultimately depended on the decisions and observations made at the operations level.

The Tables below provide measures ranging in their degree of precaution – from strictly to weakly precautious. Your choices of appropriate measures will be guided by the process outlined in Part 2, which led to decisions on *how* much precaution should be applied.

TABLE 1: Existing Operations and Activities: Precautionary Measures

GENERAL COMMENTS

Decisions made about the operational details of activities or the day-to-day running of projects can directly affect the environment. For example, in the case of forest logging operations, some 'on the ground' issues are:

- where should roads, snig tracks and log dumps be located?
- which individual trees should remain?
- what should be done to prevent erosion and ensure regeneration after logging activities?
- what can be done to guard against the transfer of plant diseases through the movement of people and machines?

Whilst many such issues may have been prescribed in management plans there may still be scope for choices that include the need for precaution.

The operational level is critical to the overall implementation of the precautionary principle. For example, without effective care taken and monitoring of operational activities by those working at the 'coalface', unpredicted environmental damage may be noticed too late.

EDUCATION

Ensure operators are educated (including contractors, ground staff etc.) about a range of issues:

- adherence to all relevant regulations, codes of practice and standards.
- adherence to Best Environmental Management Practice and Environmental Management Plans.
- auditing and monitoring (see below).

MONITORING

Monitoring may be a condition of licence or may be carried out by an organisation as part of an environmental management plan or for some other purpose. Whatever the reason for monitoring those at the operations level can play an important role. They may be well placed to detect environmental effects at an early stage and can provide a vital link within an overall environmental management system. Operators should be encouraged to feedback operational and ecological information from the 'coalface' of operations to management levels.

Formal Monitoring Programs

Develop monitoring programs as part of an environment management system or plan (see discussion on Monitoring in Table 3). This sets out a range of factors to be observed and reported periodically to ensure that dangerous limits are not exceeded. Operators at the 'coalface' should be encouraged to take a pro-active role in reporting signs of environmental change.

Incentives for Monitoring

Monitoring could be further facilitated by 'incentives' for providing information gathered from the day-to-day operations level that could assist in avoiding serious or irreversible environmental damage. Examples are: reporting sitings of endangered species in approved logging coupes; warnings of possible negative environmental effects from waste management operations of a company. Such incentives might be in the form of 'rewards' (greater than the foregone opportunity if appropriate) administered by environment protection agencies, private organisations, or by industry associations on behalf of industries. Incentives may need to be accompanied by the introduction of legislation for increased protection of employees reporting their employers (see Table 5).

Holistic Monitoring

In monitoring or surveying, examine 'processes' as well as 'things'. This means taking a more 'holistic' approach to the observations, monitoring guidelines and decisions made at the operational level. Ideally, monitoring of a range of ecosystem variables should be regular and linked to sustainability indicators that can provide early warnings and feedback. These indicators could form a list of 'triggers' for specific outcomes to be avoided at the operations level; for example, endangering habitats. Monitoring should cover pre-, current and post-operations activities.

Public Involvement

Facilitate and encourage acceptance of public involvement in monitoring and feedback programs at the operational level. Make the public feel welcome to be involved. They may have very valuable information to offer.

Note: Monitoring programs are very important, but consider their limits. Once detected through a monitoring program, effects have already occurred! This means you should not rely on monitoring procedures as your only precautionary measure. They should be used mainly as a safety mechanism for detecting possible mistakes in other management measures you have adopted.

AUDITING

Undertake environmental audits of energy and materials use in operations to determine environmental impacts. Auditing helps to determine the most effective precautionary measures at the operations level (see discussion on Auditing in Table 3).

TECHNOLOGY

Best Available Technology

For strict or strong levels of precaution, operators should use *as a minimum* the best available technologies (BAT) or techniques and practices for:

- pollution prevention and control

- waste management
- the mitigation of environmental effects
- resource harvesting activities and construction projects
- environmental rehabilitation
- other activities.

Remember that for a strict precautionary response BAT may be considered insufficient. Either BAT supplemented by demand management or not proceeding with the proposal may be a more appropriate course of action. A range of factors including the benefits from the proposal will influence this choice.

For weaker precautionary responses, best available technologies not entailing excessive cost (BATNEEC) may be appropriate. BAT means that all that is technologically feasible must be done to prevent the harmful effect, while BATNEEC means this should be done within economic reason.

Ensuring appropriate technology

While the use of BAT may be a practical precautionary measure in many situations, it is always important to ensure that technologies used are *appropriate*. This means considering a number of factors:

- What is 'best' in one situation may not be appropriate in another. Consider the context in which the technology is to be used and the goal that is to be achieved. At what expense does BAT achieve this goal? For example, if the best available gear is required by fisheries operators to ensure a more efficient catch of particular target species, will only the best funded operators be able to comply? Consider the issue of equity.
- BAT should not be used as a substitute (that is, a technological fix) for more effective measures that address the root causes of the environmental threat. For example, it is important not to confuse the efficient catch of fish using BAT with their over-exploitation. Emphasis should be on control of the total fishing effort, rather than on only the gear each contractor can use.
- Should there be measures in place to ensure the technology will be understood by those who will use it? Are the resources required for the technology to be repaired widely available?

DECISION-MAKING

In making operational decisions:

- try not to shift environmental impacts from one area to another; and,
- try to time operational activities to reduce harmful effects (for example coincide with natural cycles to produce least impact).

TABLE 2: Proposed Projects and Plans of Management: Precautionary Measures

GENERAL COMMENTS

The term project is not confined to the construction of buildings or other structures. Projects are all kinds of activities undertaken to achieve a particular result, for example, the extraction or raw materials (timber, coal), the clearing of land, the spraying of pesticides, the extraction of water for irrigation and the discharging of wastes. In some cases a project might consist of a combination of activities. Projects may be carried out by government bodies, private industry, commerce or the farming or fishing communities, or they may even be personal (for example, choosing a place to construct a house). The term 'projects' should also cover the development of new products (manufactured goods) or new technologies (biotechnologies).

With the exception of new products, which may have wide distribution and use, projects are generally site specific. They may be small to large scale (as may their effects on the environment). They may have definite starting and finishing times (such as the development of a mine), or they may be ongoing activities (such as the treatment and disposal of industrial waste in waterways). The way they are conducted or managed is determined by a *plan of management,* which may be developed by the project proponent, or by an approval authority. Projects and their plans of management involve a number of stages, including planning and design and approval by an authority if required. A range of measures can be considered in each of these stages by project proponents and decision-makers.

PRE-PROJECT DEVELOPMENT

Project Demand

First consider the demand for the project outcomes. Has this been assessed? Were the methods appropriate? Studies of 'demand' for all project activities should be made by proponents. This should consider justifications (social, economic, cultural, medical etc.) for 'using' the environment, enabling decision-makers to assess proposed projects and related plans of management in context. Demonstrated knowledge of alternative means (management regimes/technologies) for meeting the demand should also be shown. Can the possible environmental effects be justified (considering 'worst case scenario' of the best alternative)?

Project Siting

Careful selection of the project site (if appropriate) will help to avoid some of the uncertainties leading to the need for precautionary measures. Avoiding sensitive environments where impacts are more likely to be considered 'serious' is a measure that may improve the acceptability of the project.

TABLE 2

Public involvement

Ensure broad public consultation/education/negotiation at pre-development or planning stages to help identify potentially serious threats and to gain information on community perceptions regarding 'threats' and the appropriate measures.

DEVELOPMENT OF PROJECT OPTIONS

Alternatives

Assess alternative approaches before proposed developments proceed, including:

- alternative waste prevention and waste management approaches;
- demand management to spread environmental burdens as much as possible. This is particularly appropriate for managing pollution loads (eg municipal waste) and resource consumption (eg electricity use). Demand can be managed by price structures and education and may reduce the need for the proposed project;
- alternative resource and energy sources used in the project that are less environmentally damaging (undertake a life cycle assessment of materials); and,
- not proceeding with the project (that is, the do-nothing option).

Note that many of these considerations may already be expected under the regulations for project approval (for example in NSW under the 1994 Regulation to the *Environmental Planning and Assessment Act* (1979).

Natural cycles and conditions

Consider the natural environment in the design and development of the project. Attempt to replicate or mimic natural conditions in its design and implementation. For example, consider the ecological and physical effects on the whole water catchment area, the timing of project implementation to fit with natural cycles, or the siting of activities to reduce harmful effects.

Best available technology (BAT)

Use BAT to avoid or minimise damage wherever possible, particularly if the situation requires a high degree of precaution. If the best technology is too expensive, the activity should not proceed unless only a moderate or weak degree of precaution is required. In this case, use the best available technology not entailing excessive cost (BATNEEC) (see comments in Table 1).

Targets and indicators

Develop targets and sustainability goals to rank different options (eg degree of water purity or number of indicator species in an area).

Ongoing project management

Implement Best Environmental Management Practice (BEMP) and Integrated Environmental Management Programs for the management of projects. Such management systems should include clearly defined targets and unambiguous objectives to make it possible to know not only if the objectives are being met, but also what should be done to make them achievable. This requires a 'feedback loop' so that movement towards the objectives is possible (constant improvement) and 'triggers' alert managers to outcomes that should be avoided. In some circumstances, the management system itself could be trialled before it is introduced.

Public involvement

Broaden the degree of public participation in developing project options to ensure hearing of alternative plans/strategies/approaches to problems/developments.

IMPACT IDENTIFICATION AND PREDICTION

Reduce uncertainties

Wherever possible, close information gaps to reduce uncertainties. Undertake research and monitoring across a range of areas.

Baseline ecology

Reduce uncertainty about the environment itself by obtaining 'baseline' ecological information. Conduct studies of ecology as a foundation for future monitoring and appropriate responses. Take 'snapshots' of the ecosystem as basis for detecting change to establish criteria for future decision-making. Monitor for 'limits of acceptable change' as a basis for implementing adaptive or staged development (see below).

Project Impact

- Extend and refine predictive modelling by conducting targeted research and development. This will help to better understand and control impacts.
- If the project is a new technology, tools such as ecological risk assessment may help to predict impacts.
- Life cycle assessment may assist in comparing alternative components of products.
- Consider what environmental impacts have already accrued and what further impact will result from the proposal. Consider cumulative (additive) or synergistic (the impact is the more than the sum of its parts) effects.

Environmental Impact Assessment

Assessing the impact of projects on the environment before they proceed is a critical precautionary measure. Even if it is not formally required, always determine the sorts of impacts associated with proposed projects or plans of management. Assess the likely worst-case scenario impacts of the project proposal on the environment in the short, medium and long term:

Cumulative impact

Have long term, ongoing or cumulative impacts of the project including the use and disposal of associated products (and by-products) been considered? If so, should there be compensations for these?

Social impact

Who is affected? Do they have a voice in the impact assessment? Has there been broad and adequate public consultation?

TABLE 2

Land capability

What is the capability of the land? Has consideration been given to the implications of land use to adjacent or downstream areas?

State assumptions

State all assumptions underlying EIA. Is the area considered ecologically important? What ecosystem components are included in the assessment?

Data reliability

Indicate what is and what is not known to bring the limits of the impact assessment to public attention. Data reliability statements assist in the implementation of precaution and they demonstrate the credibility of the science involved and the integrity of those making the assessment.

IMPACT AVOIDANCE, MITIGATION AND REHABILITATION

Monitoring

Monitoring programs are an essential precautionary measure and should be built into the development and ongoing operation of all projects. Determine specific outcomes to be avoided (that is, 'indicators' - concentration of a particular chemical, declining number of species 'X' etc.). Whether these involve technologies that detect certain levels of a chemical or people making observations about the colour of water or the numbers of organisms in an area, monitoring can 'trigger' appropriate management responses immediately (that is, stop the activity and/or modify the activity). But monitoring is only an effective precautionary measure if the indicators provide an *early* warning. This means indicators should be set at very conservative levels to provide scope for reversing the effects of an activity.

Project implementation

One way to mitigate environmental impacts is to manage the project by step-wise adaptive management. This means using 'feedback loops' so that uncertainties are acknowledged and projects are implemented in a step by step fashion as knowledge regarding the impact becomes more certain. This safeguards against the possibility of unexpected severe future costs. In other words, it is prudent to leave some margin for error until one learns more about the consequences of a decision.

Project management

Incorporate ecological 'safety margins' in plans of management. Check there are:

- resources in reserve for 'crises'?
- buffers, barriers against ecological change?
- appropriate safe minimum standards?

Site management

Plans of management for projects should include attention to source controls on site and/ or trapping controls at site boundaries for pollutants/waste potentially leaving the site.

Mitigation technology

Use best available technology for mitigation of environmental impacts wherever possible.

Project proponents should demonstrate that technologies for environmental protection or mitigation work successfully before projects are approved

Rehabilitation

If appropriate, ensure that rehabilitation and decommissioning have been thoroughly considered in the plan of management. Again, if strict precautions are appropriate, the best available technologies, techniques or practices should be used in rehabilitation and decommissioning. Proof that these are successful in producing the required degree of environmental quality should be sought and experimental trials conducted if possible.

Offsets

Where serious impacts are possible, plans of management could include commitments to offset projects as a precautionary form of compensation. Different kinds of offsets can be offered, for example, tree planting to develop 'carbon sinks' to offset greenhouse gases generated by a power station operating on fossil fuels. Offsets may take the form of a donation of land for a nature reserve, or a contribution to landcare programs.

While they offer compensation, *offsets do not and cannot restore fully what was lost.* In nature, there can be no complete restoration and a general rule for ecological integrity is that *there is no substitute for leaving it alone.* Project offsets used as a precautionary measure should recognise this qualitative loss and compensate accordingly in the quantitative dimension. Approval conditions should take this into consideration in weighing up the costs and benefits of a proposal.

DECISION-MAKING AND APPROVAL

Decision 'rules'

Decision-makers decide whether to approve projects and if so under what conditions.

Avoid shifting problems around

Time

Ensure that any significant but uncertain environmental damage that *could* be avoided in the present, is avoided, but *not* by shifting the effects to another time. Assess projects in the context of the outcomes on future generations (this is the principle of inter-generational equity). Try to make sure they will not be left with problems created by decisions made now.

Space

The shifting of environmental impacts to other places should also be avoided (the principle of intra-generational equity). For example, it might seem to be a precautionary measure to approve a project that depends on the shipment of the toxic waste it produces to another country. But the problem would just be transferred to others who may deal with it in a more hazardous way. Decision-makers must take the responsibility to ensure that alternatives that are more precautious are fully investigated, rather than work on the 'out of sight, out of mind' principle.

Apply strict precaution first

Decision-makers should try to apply the most precautious project management options first. As more information comes to hand (perhaps after undertaking studies to reduce key uncertainties), decision-makers may then choose to ease this requirement and allow less precautious options as appropriate.

Approval considerations

Some precautionary considerations for decision-makers assessing particular projects are:

- Consideration should be given to the potential for specific design and operational features that reduce environmental impact (eg energy efficient light fittings, building aspect etc.) and decisions should be made with the aim of comfortably meeting the most stringent environmental quality objectives, such as safe minimum standards, wherever possible. Meeting all relevant regulatory requirements does not guarantee that potentially serious or irreversible environmental damage will be avoided.
- The siting of the proposed project should avoid sensitive environments.
- Assess new developments in terms of their 'sustainability' within their region or catchment. This will help to wind down dependence on unsustainably used environmental 'goods'.
- Are there commitments to monitoring and reporting in the plans of management? Monitoring of key ecological indicators before, during and after project development should be a minimum requirement for approval (see discussion below).
- Are there commitments to site rehabilitation and decommissioning (if appropriate)?
- Are there contingency plans in place that take effect should impacts prove 'unacceptable'? Such plans need to be linked to monitoring and to well understood alternative technologies or management options.

Deferred approvals

If necessary, consider deferral of project approval in the short term where there are high threats and uncertainties. As more information about the environment and possible damage to the environment comes to hand, a decision to approve the project can be made more confidently.

Interim protected areas

Seek to defer approval by declaring interim protected areas (from fisheries, mining, logging, agriculture, urban development etc.). Use increasing knowledge about the area and the proposed activities to decide whether the area should remain protected or the proposed activities can be gradually allowed. Precautionary allocation techniques at each stage of the process of land-use planning should be used. Assess proposed activities against seriousness/irreversibility criteria and the degree of scientific certainty about the likely outcome of such activities in each area. Proponents could also be required to contribute to costs of research and monitoring of interim protected areas.

Owning uncertainty in resource use

Set the permitted use or 'harvest' rate at a low or conservative level (based on general experience) where information about a resource-use activity is sparse, either because there is a new technology introduced, the location is new or the target species is little known (eg in the case of fisheries). The availability of relevant information about a resource base or species or technologies used, if linked with the easing of restrictions, would force those seeking to exploit the resource to value the collection and distribution of such data. Ideally, data collection programs could be funded directly by the resource users or through some form of levy based on the rate of harvest.

In terms of project management conditions, where only modest levels of harvest are proposed, comparatively little information is required to avoid serious or irreversible threats to the resource. When proposals for more significant harvest levels are made, the required amount of information would be correspondingly larger.

Partial (or staged) approvals

Approval can be constrained so that stages of the project known to be within the realm of complete reversibility can proceed in the short term. Later stages can be approved depending on the collection of further information. In some cases, reviews of advances in best technologies or management techniques can be undertaken to ensure projects entail more 'acceptable' levels of uncertainty or risk. This measure would be appropriate when strict precaution is required since it ensures that proponents are very confident in the soundness of their proposals.

Adaptive management

Approvals can be structured so that projects are implemented in a step-wise and adaptive fashion. Monitoring for environmental effects is constantly required through different stages of the development of the project. Feedback loops are put in strategic places and, as long as there are no signs of serious or irreversible environmental damage, the activity can continue. Wherever monitoring indicates there is evidence of such damage, activity stops and the situation is closely assessed to determine the next steps. Indicators of damage must be set at very conservative levels so that the damage which occurs before the activity stops is limited and easily reversible.

Insurance mechanisms

Assurance bonds can be required of project developers which act as a form of environmental quality insurance. Licensing for projects (see below) can be linked to these bonds, which transfer some responsibility to proponents for potentially adverse future environmental consequences. It encourages a precautionary approach by requiring investors to lodge redeemable environmental assurance bonds larger than the likely cost of recovering from the 'worst case scenario'. The size of the bond may depend on the activity, but should be at least as large as the estimated cost of recovering from serious environmental damage if it is reversible, or as much as is required to compensate for irreversible ecological losses by including *all* aspects of the loss (ecological, economic, social, cultural, scientific, etc.). Bonds can apply to all stages of projects from their development, during their operation and after their closure. They are redeemable when regulators are confident that no serious or irreversible environmental damage will occur.

Although they are not strictly an insurance mechanism, it is worth noting here that *offsets*, as discussed above, may be a condition of approval.

Licensing

Using a system of licensing is a precautionary measure based on an assessment of some knowledge of impact *before* activities go ahead. Different degrees of precaution can be incorporated in licensing arrangements. Some options are:

- refusal of license (strict precaution);
- deferral of approval (see above) with requirements for more research/knowledge to reduce key uncertainties before a licence is granted (strong precaution);
- requirement for use of best available technology (BAT) or best available evidence/

knowledge if there is a high degree of threat (moderate precaution);

- requirement for best available technology not entailing excessive cost (BATNEEC) if there is a moderate-low degree of threat (weak precaution); or
- strict monitoring requirements and a step-wise adaptive management approach (see above) to detect any harmful effects before they become serious or irreversible.

Public Involvement

Ensure as much public involvement in the decision-making process as possible. This may involve education or consultation about a particular decision and developing provisions for negotiation with the community if required. Ensure broadly based participation.

TABLE 3: Organisational Management: Precautionary Measures

GENERAL COMMENTS

There are a number of ways in which an organisation can integrate a precautionary approach into the management of its activities which may affect the environment. The internal structure and culture of an organisation and the willingness to adopt, develop and adhere to an integrated environmental management system will be very important for the successful application of the precautionary principle. At the level of organisational policies and organisational management strategies, there are a range of procedures, processes, goals or measures that can be adopted by organisations that may help to *facilitate* the implementation of the principle.

ORGANISATIONAL POLICIES

Strategic environmental assessment

Assess the environmental impact of the policies or strategies that guide the activities of the organisation. Using a process of strategic environmental assessment (SEA), assess the likely effects on the environment caused directly or indirectly by strategic, forward, financial and corporate management plans, policies or strategies. This will help decision-makers within the organisation to identify uncertainties and potential threats associated with strategic decisions and to integrate appropriate precautionary corrective measures in organisational policies.

ESD principles

Incorporate the precautionary principle and the other principles of ESD in the organisation's corporate plan or mission statement.

Long-term and holistic organisational management policies

Develop organisational policies for long-term business strategies based on environmental responsibility and take a 'holistic' approach to organisational management. Don't sweep environmental impacts 'under the carpet'. Develop policies for organisational managers and decision-makers that prohibit the shifting of environmental impacts from one place to another (intra-generational equity) or from the present to the future (inter-generational equity). For example:

- Aim for clean*er* production and to ultimately reduce pollution to harmless or zero levels (clean production, see *Pollution Prevention* below).
- Consider the effects of organisational activities beyond the boundaries of the organisation and in terms of natural boundaries (eg catchments, bio-regions). Develop the use of life-cycle assessment (LCA) in decision-making and policy development. For example, incorporate 'life-cycle' considerations in purchasing policies and in decisions involving cost-benefit analysis: *Where do the raw materials come from?*

What is the environmental impact of these materials in terms of their extraction, their transport, their use and their disposal? What are the options?

- Take on responsibility to reduce, re-use and recycle packaging materials and to take responsibility for other waste materials generated in the development of products ('cradle-to-grave' responsibility).
- Develop accounting systems that include resource use and potential environmental impacts.

Pollution Prevention

As an organisational policy, priority should be given to the use of cost-effective pollution prevention measures, such as clean production, product substitution and waste minimisation. Where cost-effective measures are not available or feasible, use best available scientific information to set standards for environmental protection (see discussion above regarding targets and indicators). The standards should: include large ecological safety margins; be revised regularly to ensure they are still appropriate; promote continuous improvement of environmental performance. Allow for flexibility in meeting these targets (that is, outcome, not process oriented), but consider the steps needed for the effective implementation of these standards within the organisation.

ORGANISATIONAL MANAGEMENT STRATEGIES

Integrated environmental management system

To implement organisational policy commitments, develop an environmental management plan (EMP) which is specific to the activities of your organisation. Such a plan will integrate the strategic organisational policies and planning strategies above and the measures discussed below to make up an *integrated* environmental management system (EMS).

Organisations are increasingly seeking accreditation for compliance with the ISO 14000 Standard for EMSs, to demonstrate their commitment to improved environmental performance. Figure 8 indicates possible components of such an EMS and how an EMS leads to continuous improvement for an organisation. Some of these EMS components are discussed below.

Environmental auditing

Through an environmental audit, the activities of an organisation and their known or potential environmental effects can be systematically listed and re-evaluated. This enables you to take action to stop things which you are *certain* cause serious or irreversible environmental damage (prevention) and to stop those which *might* cause serious or irreversible environmental damage (precaution). Regular auditing helps:

- evaluate whether an organisation has all of the necessary environmental permits and licences (see *Due Diligence* below) and indicate applicable variances or exceptions from regulation;
- evaluate potential liabilities (eg contaminated land);
- protect people in terms of environmental safety and legal liability of employees;
- lessen government's desire to inspect and/or make inspections more manageable;
- deter prosecutions, avoid financial problems;
- in the selection of business contractors;

- develop a list of environmental priorities and pinpoint where and how environmental performance can be improved through either changes in design or procedures and whether money can be saved; and
- identify the areas where the precautionary principle is applicable and help the organisation to be proactive about applying precautionary measures.

Note: *One of the aims of the Protection of the Environment Operations Bill is to ensure that information which becomes apparent to an organisation through auditing cannot be used as evidence against it in prosecution. This law will encourage more organisations to undertake voluntary environmental audits.*

Environmental due diligence

Some organisations are required to demonstrate environmental *due diligence* under different State laws (for example, in NSW this is a requirement under the *Environmental Offences and Penalties Act* 1989). Satisfying due diligence requirements can be linked to auditing (above) and the development of an integrated environmental management plan or system in general. For example, Sydney Water has defined environmental due diligence as ‘using all reasonable foresight and care in planning and carrying out our activities so that the harmful impact on the environment is minimised or prevented’. One of the aims of the environmental due diligence management plan they have developed is to ‘establish systematic processes to continuously identify environmental hazards involving current or anticipated events, activities, or incidents that may result in environmental harm’. This allows the organisation to continuously improve its environmental performance, demonstrate due diligence and to apply the precautionary principle wherever appropriate.

Targets and Indicators

Develop a set of sustainability targets to be achieved and outcomes to be avoided. Use a system of indicators to measure how well the organisation is performing against these targets. Link these indicators to the EMS in a way that sets out or ‘triggers’ particular actions for a range of different decision-makers in the organisation. An example would be reporting to project managers that a particular level of pollutants had been observed at the operational level. If the level is particularly high, project managers would be required to shut the activity down immediately. If a lower level was indicated, further investigation might be required as the first step and further action taken on this basis. Indicators need to be constantly revised in the light of new information (see R&D below).

Monitoring

Monitoring is critical for improving an organisation’s environmental performance.

Projects and operations

Monitoring programs should be developed to provide early warning about environmental damage arising from specific projects and activities conducted by the organisation (see Table 2: Projects and Plans of Management). Monitoring helps to meet ‘safe minimum standards’ and to contribute more broadly to the development of sustainability indicators and for a baseline to assist in continuous improvement.

Precautionary measures

Monitoring should also take place to assess the effectiveness of any precautionary control measures put in place. If these do not prove effective or have not been successfully implemented, then appropriate corrective measures can be undertaken.

Environmental management system

Combined with regular auditing and the development of a system of indicators (see below and Figure 8), monitor and assess the effectiveness of the environmental management system as a whole. *Are there areas in the operation of the organisation that can be improved in order to improve overall environmental performance?* This could be part of environmental auditing for the organisation as a whole and linked to reporting mechanisms to demonstrate no serious or irreversible effects arising from existing operations.

Environmental and continuous improvement programs

Organisations can undertake programs for continuous improvement as an indirect precautionary measure. This means monitoring or auditing current environmental performance and improving this performance across a range of indicators. Continuous improvement involves a policy commitment to regular and thorough updating of all operational and management activities. Steps to implement policies for improved performance can be undertaken by both industry and government in this direction.

Industry

Individual organisations may take the initiative in developing improved environmental performance (for example, through adopting an environmental management system based on the ISO 14000 series and using this to implement continuous environmental improvement), or industry associations may take the initiative by requiring their members to improve environmental performance. For example, a requirement for membership of the Plastics and Chemicals Industries Association is that businesses must comply with strict environmental standards and Codes of Practice as a part of the association's Responsible Care Program. Such programs are increasingly common.

Government

State governments can implement policies for the provision of guidelines, advice and the development of incentives for industry to undertake programs of environmental improvement. For example, the South Australian Environment Protection Authority negotiates formalised agreements with industry, with priorities identified, proposed action outlined and time frames set for environmental improvements. Commitments to undertaking these environment improvement programs can be conditions of licences. This approach provides greater certainty for the community, environment and industry, while promoting best practice environmental management among industry.

Dealing cautiously with hazardous substances

Where organisational activities involve potentially hazardous or toxic substances, ensure that there is a management plan for dealing with these substances. This plan should be widely understood within the organisation and should contain:

- a list of substances known to be hazardous or toxic, or thought to be hazardous or toxic, clearly linked to the stages or processes in which they are involved. This list should be easily understood, regularly updated and changes should be well publicised within the organisation and linked to occupational health and safety requirements.
- explanation of the possible effects of these substances on human and environmental health.

- best-known procedures for dealing as safely as possible with these substances. This should be widely understood within the organisation, not just at the operational or handling level.
- the required responses for dealing with these substances in emergency situations.

Research and development activities

Conduct, sponsor or promote broad-scale environmental research to contribute to knowledge about the natural environment. Conduct research and development for the ongoing improvement of technologies used for environmental protection or which contribute to the goal of ecologically sustainable development. Where scientific uncertainties arise as to the sufficiency of standards or targets, use targeted research to reduce uncertainties if possible and review and improve control measures applicable by the organisation.

Education and employee awareness

Organisations should educate employees at all decision-making levels about a range of issues relevant to a precautionary approach. Aim to develop their capacity to be 'agents of change' in moving the organisation towards stated environmental goals.

Responsibility for precaution

Ideally, an organisation should work the principle into its everyday activities at all levels by making sure everyone understands it in relation their own decisions. Alternatively, an organisation could ensure there is someone (or perhaps a few people) responsible for 'handling' the precautionary principle and incorporating it in decision-making, perhaps as part of an environmental management program or in relation to other company or departmental decision-making. *If staff at all levels of an organisation do not appreciate the relationship between their decisions at work and the environment, the precautionary principle will be difficult to apply effectively.*

- Aim to ensure no decisions are made without consideration of the need for precautionary measures. This means making sure there is no excuse for someone saying 'I thought that was your job' or 'Isn't that the boss's job?'
- Consider running seminars on the precautionary principle to inform and involve employees, generate organisational approaches and identify appropriate precautionary measures.

Life-cycle assessment

Education about life-cycle assessments should be a goal within an organisation. All levels of decision-makers within different types of organisations (public and private) should understand LCA in relation to activities they are involved with, including its benefits and its limitations.

ESD

Educate employees about ESD and emphasise the potential impacts of the organisation beyond its boundaries through its products and processes.

Involvement of employees

Organisations should recognise as an important asset, workers with knowledge gained through years of experience at the operational level. This recognition should be given a high profile and coupled to an environmental management system that encourages workers at this level to be involved in application of the precautionary principle. For example, beyond merely following orders or guidelines, a system of incentives could be developed for acting to stop and report as soon as possible potentially serious or irreversible damage that appears at the operational level. Involve employees in development of the organisation's EMS.

Involvement of the public

Ensure 'transparency' of strategic decisions by involving the public in decision-making as much as possible. Ensure open communication with the public. Consider setting up public education campaigns, Public Advisory Committees and Responsible Care programs if appropriate.

Figure 8: Facilitating the Precautionary Principle: Integrating an EMS with organisational policy and management strategies for continuous environmental performance improvement

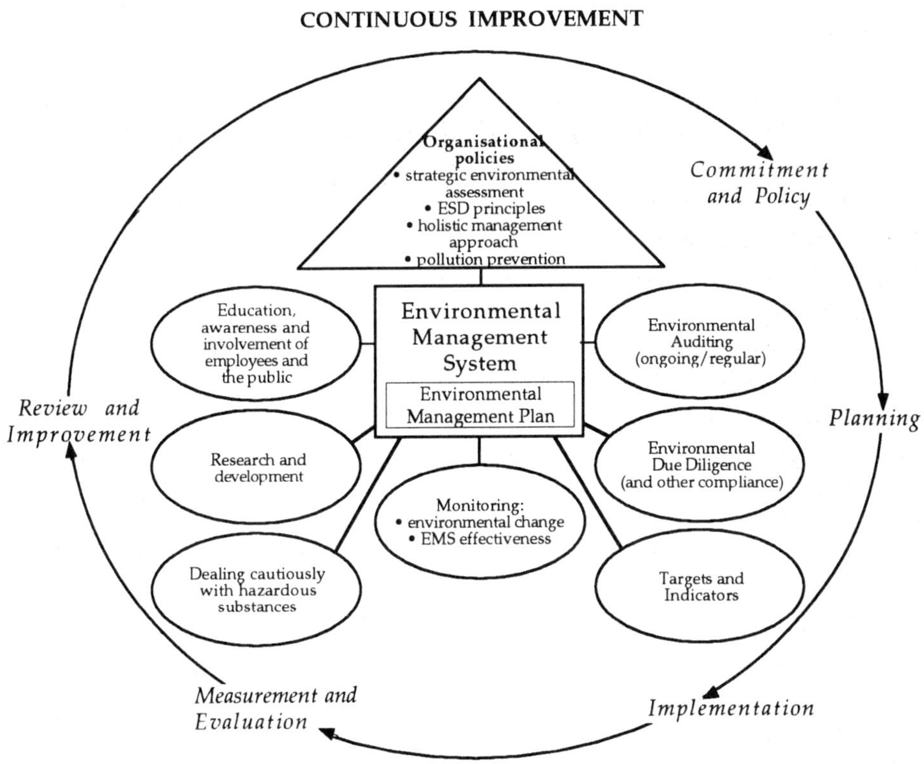

The organisation should adopt policies that focus and commit it to the application of the precautionary principle as part of better environmental performance (for example, by incorporating the precautionary principle and the other principles of ESD in the mission statement). A range of organisational management strategies can be adopted and ideally, these will be linked to produce an integrated environmental management system: all organisations should be aiming for continuous improvement.

TABLE 4: Adminstrative Procedures: Precautionary Measures

GENERAL COMMENTS

Responsibility for administering decision-making about activities affecting the environment lies with various government authorities or departments. These 'determining authorities' follow particular *assessment and decision-making procedures* to determine whether approval should be granted and what sorts of licensing conditions are applicable. They can also be involved in developing policies and laws about how these procedures work (see Tables 5 and 6). But the overall set of *administrative arrangements* or relationships between different government authorities and departments, industry and the public also affects the way individual decisions are made. Changes to some of these arrangements might facilitate the implementation of the precautionary principle.

ASSESSMENT AND DECISION-MAKING PROCEDURES

Economic instruments

Undertake a critical evaluation of the economic tools used to assess proposed and pre-existing developments and incorporate improved pricing mechanisms and natural resources accounting in administrative and fiscal decision-making. Assessments should be based on full environmental, social, and inter- and intra-generational costs.

Impact assessment

Many 'projects' are required to undergo environmental impact assessment (EIA) and submit an environmental impact statement (EIS) before they go ahead. Assessing the potential environmental impact of projects before they are approved is a precautionary procedure. With the information produced by an EIS, further precautionary measures may also be put in place.

Justification of activities

Studies of the 'demand' for all projects and activities planned should be taken into consideration by decision-makers (see Table 2).

What is assessed?

Build precaution into the procedure for classifying or identifying activities that require EIA/EIS in the first place. For example, review definitions for activities classified as 'designated developments' to ensure that all potentially harmful activities require impact assessment before approval. In addition to the assessment of 'projects', such as the construction of buildings and other structures, assessment should be made of the environmental impacts of:

- technologies, techniques, processes and products; and,
- policies, strategies, plans and programs.

TABLE 4

Types of assessment

Emphasise other types of assessment that could be in the impact assessment process:

- inter-generational impact assessment (what are the environmental impacts on future generations?);
- social impact assessment (SIA — the social impact of activities); and,
- cumulative impact assessment (CIA — assessment of the accumulated impacts of an activity and previous activities).

Data reliability

Increase the degree of 'transparency' of EIA processes by requiring that uncertainties are identified and a data reliability statement is given.

Selection of consultants

EISs are generally prepared by consultants selected by the proponent. But this practice helps to entrench biases in favour of the proponents since it is in the interests of the consulting party to contract future work. Redressing this bias will require the development of a consultant selection process. This will help to implement the precautionary principle by improving the likelihood of bringing all uncertainties and impacts to the fore in impact assessments.

Inquiries

Include the precautionary principle in the terms of reference for inquiries to ensure that it is given due recognition in strategic decision-making and at other levels affected by such inquiries.

Environmental Empowerment

There are many uncertainties and subjectivities in deciding what is 'serious' or what is 'irreversible'. This means that there is a need for broad input into judging the right response for dealing with activities with possibly significant but uncertain environmental possibilities. Unless regulatory decision-making procedures provide meaningful points of access for these subjective judgements, applications of the precautionary principle will be limited or ineffective. A number of measures can be taken to modify assessment and decision-making procedures to empower the public in environmental matters and increase the breadth and depth of their involvement.

Information

Make information as publicly available as possible.

- Ensure freedom of access to environmental information.
- Panels for the production of statements on controversial issues can assist in the development of public understanding.
- Introduce procedures at a range of administrative levels based on a principle regarding the sharing of information relevant to environmental protection. This principle ought to override 'commercial-in-confidence' agreements if it can be shown that the prevention of potentially significant environmental damage outweighs the significance of the confidentiality agreement. Penalties may be appropriate to accompany violations of the principle of information sharing wherever it can be shown that relevant information has been deliberately withheld. This may be tied in with the introduction of legislation that brings about enhanced protection for employees who 'blow the whistle' against their employer in relation to environmental offences (see Table 6).

Transparent decision-making

Make decision-making processes transparent and ensure effective and timely public consultation. This may include open negotiation and ensuring that adequate community representation exists.

Non-technical summmaries

Proponents and the relevant determining authorities (perhaps through an agreed process or 'partnership') should publicise intended activities in non-technical terms before decision-making processes commence. Though a non-technical summary is generally required in preparing environmental impact statements, this is not really circulated beyond the EIS itself. The likely economic and social benefits and the possible ecological and social costs (determined by environmental risk assessments) should be summarised and made very widely available (eg through the print media) prior to the proponent's submission of the EIS. This assists in the implementation of the precautionary principle by increasing the possibility of greater public involvement at the scoping stage of the EIS process as well as through public submissions to EISs and public inquiries. It allows the public to better know whether they have an 'interest' in the issue earlier than is currently the case and increases the transparency of decision-making.

'Stop orders'

In various regulations and administrative processes, the ability for a wide variety of 'stakeholders' to issue a 'stop-order' on a range of activities could be introduced. This procedure would only take effect (through the appropriate determining authority) if there were 'reasonable and probable grounds' that serious or irreversible environmental damage was occurring or was likely to occur.

Deferred approvals

A precautionary measure at the level of administrative procedure is to defer approvals if there is insufficient information or concerns about significant environmental threats. For example, amendments to Part 4 of the NSW *Environment Planning and Assessment Act* 1979 allow the Department of Urban Affairs and Planning to defer approvals when assessing proposals. This legislation makes it possible for the Department to defer an approval if it is judged that there should be some proof (through a trial) that a proposal will satisfy concerns about ecological sustainability.

Insurance mechanisms

Mechanisms that provide some form of 'bond' or insurance against environmental damage (eg financial assurances) can be used as a precautionary measure by decision-makers if they are incorporated into the administrative procedures they are required to follow. These assurances might be required as a condition of the approval of some projects or operations. They are redeemable when it is certain that no irreversible damage has or will take place. This sort of mechanism makes the proponent responsible for precaution by making them ensure there won't be any (avoidable) damage (see Table 2).

Standards and codes of practice

Ensure that administrative procedures are based on standards or codes of practice that: do not impede the use of recycled materials; require inbuilt ecological safety margins as part of best management practice.

TABLE 4

ADMINISTRATIVE ARRANGEMENTS

General comments

Consider what measures might be taken to adjust the administrative arrangements for decision-making to help implement the precautionary principle.

Relationships

- Are the bodies involved in administering decision-making about the activity appropriate?
- Should other bodies be involved as well to improve the balance of interests, or, are there already too many authorities concerned?
- Do the relevant determining authorities work co-operatively with each other in decision-making to seek a mutually agreeable outcome, or are decisions produced as compromises through more antagonistic processes?
- Is there mostly a sense of co-operation or antagonism between the determining authorities and development proponents?
- Can the 'culture' of either or both parties be altered to improve co-operation in reaching outcomes? For example, proponents may become more proactive in building precautionary measures into their activities and demonstrating how they have applied the precautionary principle. Determining authorities might produce and publicise guidance on some of the measures that proponents could take to be more proactive and precautionary in relation to these activities.

Consistency

Where there is a lack a co-operation between proponents and various determining authorities, this is likely to be because of inconsistencies in the overall administrative arrangements.

Are there different regulations affecting the administration of particular activities that cause conflict between ecological and economic considerations? For example, there are some regulations that allow (and even encourage) 'over-development' or 'over-exploitation' in particular resource sectors (such as construction, mining, forestry, fisheries and agriculture). One authority may be able to approve 'explosive' development while another is required to assess it on the basis of environmental effects. Where there are inconsistencies in the 'rules' governing decision-making authorities, this causes greater stress on the environment and the need to apply stricter precautionary measures at the project and operational levels. It is also a source of frustration for developers and creates antagonism between different arms of the same government.

Conflicting administrative procedures need to be over-hauled. Review the relevant rules and regulations governing investments in resource 'development' activities and examine inter-governmental and inter-departmental relations. The overall system of regulations needs to be redesigned to encourage resource development investments that are ecologically sustainable rather than those which simply produce short-term economic gain.

Resources

- Are resources sufficient to ensure effective administration and decision-making?
- Are the available personnel appropriately trained?

- Can research be funded where scientific uncertainties might prevent profitable activities from proceeding?
- Could arrangements be made for co-operative funding of ongoing research aimed at reducing uncertainty?

Regulatory powers

- Is the overall system of decision-making flexible enough to produce appropriate precautionary responses in different situations?
- Are new powers needed to improve the ability of decision-making authorities to apply the precautionary principle appropriately?
- Do the regulations that determine the charter of determining authorities directly require them to consider the precautionary principle? For example, under Schedule 2 of the *Environmental Planning and Assessment Regulation* 1994 all projects which require an EIS must be justified in terms of the principles of ESD. The precautionary principle is specifically mentioned as one of these principles (see also discussion of 'enabling legislation' in Table 5).

TABLE 5: Legislation and Law Reform: Precautionary Measures

GENERAL COMMENTS

The precautionary principle can be directly implemented or indirectly facilitated by a range of legal measures. Some of these measures already exist, but some laws may need to be reformed or new laws introduced. ***Direct precautionary legal measures*** either require behaviours that adhere to the principle, or prohibit those that do not adhere, or both. For example, laws can prohibit or regulate the conduct of particular kinds of actions or activities, technologies or techniques, substances, projects, products or processes. ***Indirect precautionary legal measures*** are those that create an environment of incentives and disincentives which will tend to generate behavioural adherence to the precautionary principle.

DIRECT PRECAUTIONARY MEASURES

Enabling legislation

Legislation which leads to the establishment of environmental regulators or management agencies can require these bodies to apply the precautionary principle in decision-making. In some legislation it is directly mentioned as a guiding principle.

Direct prohibition

Through prohibition of specific activities or types of activities, potentially serious or irreversible environmental damage may be avoided, whether there is scientific proof that such damage will occur or not.

Reverse listing

Reverse listing is an example of a strongly precautionary measure that can be applied at the legal, policy or administrative levels to prohibit particular substances or activities. Rather than produce a list of substances that are legally prohibited for sale or dumping for example, because they are known to be hazardous (a preventative measure), a reverse list contains only those substances believed to be safe. This means that all substances are assumed to be harmful to the environment until there is good evidence that they are not, giving the environment rather than potential pollutants the benefit of the doubt. The long-term effect of this listing is that the onus is then on the developers, users and disposers of particular substances to prove that they should be on the list. Reverse lists should be connected to liability laws (see below).

Regulatory standards

Specific activities or categories of activities with the potential to pose threats of serious or irreversible environmental damage can be regulated by legal requirement to adhere to particular standards. Such a requirement is a precautious measure when applied in the absence of scientific proof that such damage will occur. The degree of precaution depends on the assumptions incorporated in the standard. Adherence to the standard may also be linked to liability and due diligence regulations.

INDIRECT PRECAUTIONARY MEASURES

Procedural requirements

Proponents of some types of activities are required by law to follow precautionary procedures such as submitting detailed environmental impact statements or statements of environmental effects about current or intended activities. These procedures are indirect measures for applying the precautionary principle. Changes to these laws could facilitate the application of the precautionary principle. Some examples are:

- requiring environmental standards used in different sectors to incorporate demonstrable environmental 'safety margins'.
- requiring monitoring before, during and after all activities potentially affecting the environment. This requirement could be linked to insurance/assurance bonds (see Tables 2 and 4) so that proponents that fail to undertake appropriate monitoring forfeit their bond where EIA predictions prove to be inaccurate.
- requiring regular environmental performance reporting to the appropriate determining authorities.
- requiring action to be taken immediately where there are clear cases (that is, 'reasonable and probable' grounds) for application of the precautionary principle and making accompanying provisions for penalties for failing to act (see also Liability below).
- incorporating a process for guaranteeing a diversity and balance of public and expert inputs in a range of decision-making processes (see also Environmental Empowerment in Table 4 and Public Involvement below).

Policy measures

Introduce legislation to give a legal basis to a range of policy measures (such as those mentioned in Table 6). For example, laws could be introduced to give a basis for increasing 'cradle-to-grave' responsibilities of manufacturers or others generating hazardous or toxic wastes or large amounts solid waste.

Public involvement

Laws that allow broad public inputs in decision-making and access to information and the courts assist the implementation of the precautionary principle because citizens' groups can act early in the decision-making process. A number of laws empowering the public already exist at various levels. Laws can be amended or introduced to empower the public to:

- act to ensure that decision-makers are fulfilling legislative obligations to use the precautionary principle to guide decisions.
- assist in the determination of appropriate precautionary measures by written submission or by means of a representative panel.

Following Ontario's Environmental Bill of Rights 1993, some examples of legal measures that broaden public inputs are:

- allowing those, who believe that an existing policy, Act, regulation or instrument should be amended in order to protect the environment, to apply to an Environmental Commissioner for a review by the appropriate Minister. Such laws could also allow for the right to make an application for the introduction, adoption, passing or issuing of new policies, Acts, regulations or instruments.
- creating a registry containing a data base of policies, acts, regulations and instruments affecting the environment - accessible by modem or from a public library - and granting public rights to comment on, call for a review of, or appeal against the items posted on this registry (though it may be appropriate that it remain the responsibility of the Minister to decide whether or not to review requests).
- the introduction of new public rights of appeal following decision-making.
- allowing for application to investigate the possible contravention of laws, policies and regulations by providing new public rights of action to prosecute polluters and to protect public resources. This would involve:
  — wider rules of standing and rights to sue those suspected of breaking existing laws and causing significant harm to a public resource, and,
  — enhanced protection for employees who 'blow the whistle' against their employer in relation to environmental offences.

Liability

Different States have liability laws to deter wilful or negligent environmental damage by individuals and corporations (eg the NSW *Environmental Offences and Penalties Act* 1989). So as well as being a mechanism for encouraging precautionary behaviour by deterring environmental harm, a liability regime can also function as a mechanism through which to pay for ecological debt. There are two types of liability regimes — strict and absolute liability:

Strict liability regime

This approach to criminal liability requires only proof that an act took place, but not that there was an intention to carry it out. The rationale for strict liability is that there are many activities which are very dangerous and that a strict offence will act as a very clear deterrence. In order to succeed in a defence the accused must have exercised due diligence in the prevention of an offence, rather than by undertaking remediation following an offence. They are able in some circumstances to have made an honest or reasonable mistake.

Absolute liability regime

Under the absolute liability regime, offences have no defence of an honest or reasonable mistake. This is a much more strongly precautionary instrument than the strict liability regime, because prosecution need prove only that the act itself occurred, regardless of due diligence.

The extension of these liability regimes to be effective over a broader range of activities and environmental standards would assist in facilitating the application of the precautionary principle.

Note: *While liability regimes can act as a clear deterrence for proponents of some activities, this should not be relied upon as a means to prevent or regulate activities posing potential but uncertain threats of environmental damage. Ultimately, the most precautionary regulatory regimes are those that involve scrutinising proposals* before *they are allowed.*

TABLE 6: Policy, Strategy and Planning: Precautionary Measures

GENERAL COMMENTS

Policies, strategies and broad planning decisions are made by government departments and other bodies at the national, state and local government levels. At this level of strategic or long-term decision-making, the environment can be affected *directly* when policies concern land-use issues (eg reserving a certain percentage of land for forestry activities), or *indirectly*, when they encourage government authorities and departments, industry and others to behave in particular ways. Policy makers and planners can encourage precautionary behaviour through a range of measures.

Note: *While there are many long-term measures that would facilitate the application of the precautionary principle, public pressure is required for governments to be empowered to take the appropriate longer-term actions immediately.*

DEMAND

Determine 'real' demand

Encourage and support demand studies for a range of activities to determine 'real' resource requirements within a planning area to help manage demand. For example:

- How much electricity do we really need?
- Often things are supply driven not demand driven. Once supply is available then there is a big push to sell, which wastes resources.

Demand management

Develop policies and strategies which aim to reduce dependence on increased supply/use of resources through flexible pricing (that is seasonal or scarcity pricing) and other market mechanisms and restrictions. Institute demand management across a range of goods and services to ensure their optimal use. For example, shift the emphasis of resource extraction activities to the 'shoulders' rather than the 'troughs' of supply (particularly where storage is a viable option). In NSW for example, this could be achieved by using the conditions of approval under the *Environmental Planning and Assessment Act* 1979.

DEALING WITH UNCERTAINTY

Uncertain environments

Implement a policy or a range of strategies to reduce basic ecological knowledge gaps and other uncertainties in key areas that prevent overall strategic management and planning from occurring on a sound basis. For example, the EPA's program of water quality monitoring in the Hawkesbury Nepean is a long-term measure that will help to implement appropriate decisions in relation to specific future proposals and developments. With a more complete information basis, more ecologically appropriate developments will be made possible and decision-makers assessing particular proposals will better appreciate the nature of the uncertainties involved. For particular regions, catchments or types of environments, such strategies will be best developed or co-ordinated between a range of regulatory bodies and the major users of environmental 'goods'.

Uncertain activities

Develop policy measures to set out conditions of approval for activities that are inherently uncertain, such as the use of new or untested technologies or previously unexploited resources. Though individual cases should be assessed on their merit, categories of activities about which there is very little certainty could be developed for this purpose. As uncertainties about certain types of activity are reduced through conservative experimental trials, conditions for approval could be made more lenient. A system of incentives such as fees and assurances could be used to put the onus for ensuring confidence about an activity onto its proponents. In this way they will be encouraged to reduce uncertainties and/or to develop alternatives known to be not significantly threatening. Minimum information requirements could be established for different categories of activities. Data collection plans could be required by decision-makers, particularly when new and exploratory activities are proposed and there is very little background information.

LAND USE

General planning instruments

Upgrade land-use planning processes to ensure that the precautionary principle is included at each stage, from strategic broad zoning, to individual development proposals. This means attention must be paid to uncertainties and to being more cautious or conservative where uncertainties are high. If uncertainties (either about the environment or about the sorts of impacts proposed) are reduced through the development of new knowledge, the planning structure should be flexible enough to accommodate this in planning outcomes. This will involve ensuring that knowledge about risks or impacts of broad level land-use activities is continually up to date and used to reassess decisions if appropriate. This will act as an incentive for those proposing to change the environment to reduce uncertainty (see above). Interim protected areas and deferred approvals may be useful mechanisms to achieve this goal. Such mechanisms would also help to lessen concerns in some sectors about the 'locking away' of areas of land.

State Environmental Planning Policies & Regional Environment Plans

SEPPs and REPs are government planning instruments used in some States in Australia that can be used to ensure that activities in environmentally sensitive areas are brought into the planning system at the regional and State levels. This allows an appropriate level of assessment to be undertaken in specific cases, enabling decision-makers to apply the precautionary principle. They provide protection to prevent or minimise potential irreversible damage resulting from a proposal, whether there is scientific certainty about it or not.

Land capability and suitability

Land should be zoned on a conservative assessment of land capability and suitability. Ensure that in land capability assessments, consideration is given to the implications of land-use to adjacent or downstream areas.

Conservative development policies

For the approval of new developments or the ongoing licensing of existing activities, decision-making should be guided by policies that use ecologically conservative estimates, regardless of the existence of any evidence of harm. A good example of this is a policy of conservative stocking by the Pastoral Board of the South Australian Department of Environment and Natural Resources. In exercising its jurisdiction over the grazing limits on lands subject to pastoral lease in SA, the Board has introduced a policy based on new stocking maximums and offers rental incentives and rebates to lessees who stock in a conservative manner.

Buffers

Develop strategies or policies to facilitate the provision of buffers or 'green belts' via zoning or land acquisition (for example, land adjacent to National Parks or belts between areas).

Zoning

Rezone areas to exclude potentially serious or irreversible activities from sensitive environments.

Planning boundaries

Planning boundaries are arbitrary political boundaries, whereas catchment boundaries are natural boundaries. Ultimately, the administration of planning and environment related departments should be along the same lines, reflecting the ecological rather than current planning boundaries. This is a very long-term policy measure which would facilitate the implementation of precaution.

ECONOMIC INCENTIVES AND INSTRUMENTS

General

Use economic instruments that provide incentives and disincentives to produce desired precautionary behaviour. This means economically rewarding or encouraging 'good' environmental behaviour and punishing or discouraging 'bad' environmental behaviour.

Note: *Careful attention should be paid to the issue of equity when developing and using economic instruments to encourage precaution. Unforseen negative economic and possibly social consequences may result for those groups in society that can least afford to alter their behaviour.*

Resource pricing

Resources should be priced to reflect *all* costs in their extraction and consumption. This means environmental costs as well as social and economic costs to the current and future generations should be considered. Policies could be implemented to charge levies for the extraction of resources, used in part to compensate the 'losers' and in part to fund research and monitoring related to environmental protection and more sustainable alternatives (see Improved pricing mechanisms below).

Improved pricing mechanisms

Undertake a critical evaluation of the economic tools used to assess proposed and pre-existing developments. Assessments should be based on full environmental, social, and inter- and intra-generational costs, which would be supported or complemented by a number of specific legal and policy changes. Some of these are:

- use discount rates which benefit environmental protection.
- develop and implement a national system of natural resources accounting (see below).
- change taxation and/or fiscal policy to encourage the maintenance of environmental quality and natural capital. This means not assuming natural 'assets' are all substitutable through the development of 'built capital'.
- incorporate disposal levies in the 'cost' of products, particularly 'disposable' products.
- develop policies and legal frameworks that require industry to take back waste from packaging and unusable by-products and for industry to adopt 'cradle-to-grave' responsibility for the disposal of their products.
- develop a sustainability tax on returns from financial investment strategies that are deemed 'unsustainable' (see Investment Strategies below). This would include all investments in the fossil fuel industry and in other non-renewable resources industries. It could be redistributed in a number of ways, including to funds to redress intra- and inter-generational equity imbalances and to a 'sustainability dividend'.
- develop a sustainability dividend (from the sustainability tax) for investors in sustainable technologies and resource development activities and for ecologically sustainable development (ESD) related research and development (R&D) activities (see below).

Incentives for sustainable product design

Introduce incentives for design, development and marketing of ecologically sustainable goods/services. These incentives include direct economic support and subsidies, as well as economic disincentives for the production of ecologically unsustainable goods and services. Policy measures using these sorts of incentives would be supported by a national system for eco-rating and eco-labelling of all goods and services.

Research for ecological sustainability

Governments should encourage major universities and government and private research and development institutions to undertake research that will contribute to the goal of ecologically sustainable development (ESD). This would involve the development and implementation of a system for classifying research that is relevant to ESD and a system for determining the allocation of funds in a way that does not distort the distribution of socio-economic objectives in existing research funding programs.

Purchasing policies

The purchasing policies of all arms of government should be changed to reflect a proactive and precautionary government. This means undertaking a thorough audit of existing purchasing practices using tools such as life-cycle assessment and overhauling unsustainable practices. This would influence the market and encourage others to follow suit. Such a purchasing policy would be supported or complemented by a national system for eco-rating and eco-labelling of all goods and services.

Natural resources accounting

Natural resources accounting is a tool that would assist in implementing the precautionary principle at a range of decision-making levels, particularly policy and planning. As long as factors such as the clean up of polluted waterways are counted as positive economic outcomes which contribute to the GDP, the national accounts will not reflect the negative environmental effects of some sectors of the economy. A system of natural resources accounting would help to give the public and decision-makers a clearer picture of the nature and quality of national economic activity. It would act as a public education campaign about the real or hidden environmental 'costs' inherent in the economic system and would also provide a useful 'benchmark' for developing more ecologically sound macro-economic policies.

Resource use policies

Policy instruments or incentives could be introduced to encourage more ecologically sustainable resource uses. This measure would be particularly useful as a condition applied to resource extraction or harvest activities that require the guidance of the precautionary principle. So despite possible threats of serious or irreversible damage, some resource developments considered economically or socially very important might be approved if there are agreements concerning the use of these resources. For example, such a policy would restrict the use of forest timbers for disposable, short-term woodchip products and favour more durable uses such as for housing projects. Such a policy would require thorough life-cycle assessments.

Investment strategies

Develop indicators of the degree of ecological 'responsibility' or sustainability of particular financial investment strategies. These indicators could be used to make clear to the public and decision-makers the connections between investments (private and public) and potential environmental damage. This would better enable precautionary investment strategies by individuals and it would tend to make governments more precautious and ecologically accountable in their decision-making.

REFERENCES AND FURTHER READING

Beder, S. (1996), *The Nature of Sustainable Development*, 2nd Edition, Scribe Publications Pty Ltd, Victoria, Australia.

Butteriss, C. and Sinden J. (1994), 'Rural Waste Management New South Wales: An Economic Application of the Precautionary Principle', *Australian Journal of Environmental Management*, Vol. 1, pp. 156-74.

Cameron, J. and Abouchar, J. (1991) 'The Precautionary Principle: A Fundamental Principle of Law and Policy for the Protection of the Global Environment', *Boston College International and Comparative Law Review*, Vol. 14, pp.1-27.

Christie, E. (1993), 'The Eternal Triangle: The Biodiversity Convention, Endangered Species Legislation and the Precautionary Principle', *Environment and Planning Law Journal*, Vol. 10, No. 6, December 1993, pp. 470-85.

Commonwealth of Australia (1992), *National Strategy for Ecologically Sustainable Development*, Australian Government Publishing Service, Canberra, Australia.

Commonwealth EPA (1994), Review of Commonwealth Environmental Impact Assessment. Executive Summaries.

Dovers, S. (1995), 'A Framework for Scaling and Framing Policy Problems in Sustainability', *Ecological Economics*, Vol. 12, pp. 93-106.

The Environment Council (1993), *The Precautionary Principle: Making it Work in Practice*, Workshop Report, 24 November 1993.

Freestone, D. and Hey, E. (eds) (1996), *The Precautionary Principle and International Law: the challenge of implementation*, Kluwer Law International, Cambridge.

Garcia, S.M. (1994), 'The Precautionary Principle: its Implications in Capture Fisheries Management', *Ocean and Coastal Management*, 22 (1994), pp. 99-125.

Gilpin, A. (1995), *Environmental Impact Assessment*, Cambridge University Press, Melbourne, Australia.

Hey, E. (1992), 'The Precautionary Concept in Environmental Policy and Law: Institutionalising Caution', *Georgetown International Environmental Law Review*, Vol. 4, pp. 308-18.

Hickey, J. and Walker, V. (1995), 'Refining the Precautionary Principle in International Environmental Law', *Virginia Environmental Law Journal*, Vol. 14, No. 3, Spring 1995, pp. 424-54.

Institute of Environmental Studies (IES) (1993), The Precautionary Principle – Conference Papers, 20-21 September 1993, University of New South Wales, Sydney, Australia.

Intergovernmental Agreement on the Environment (1992) May 1992, Heads of Government in Australia.

O'Riordan, T. and Rayner, S. (1991), 'Risk Management for Global Environmental Change', *Global Environmental Change*, Vol. 1, No. 2, pp. 91-108.

O'Riordan, T. (1994), 'The Precaution Principle in Environmental Management', in Ayres, R. and Simonis, U.E. (eds), *Industrial Metabolism: restructuring for sustainable development*, UN University Press.

O'Riordan, T. and Cameron J. (eds), *Interpreting the Precautionary Principle*, Earthscan Publications, London.

Rogers, M. and De Lacy, T. (1995), 'What's new about BCA and risk analysis: its application to the precautionary principle', Proceedings: Australia New Zealand Society for Ecological Economics Conference, Coffs Harbour, NSW, November 1995, pp. 513-20.

Seldner, B. and Cothrel, J. (1994), *Environmental Decision Making for Engineering and Business Managers*, McGraw-Hill, Sydney.

World Commission on Environment and Development (WCED) (1987), *Our Common Future*, Oxford University Press, Melbourne.

Wynne, B. (1992), 'Uncertainty and Environmental Learning', *Global Environmental Change*, Vol. 3, pp. 111-27.

Young, M. (1993), *For Our Children's Children: Some Practical Implications of Inter-Generational Equity and the Precautionary Principle*, Occasional Publication No. 6, Resource Assessment Commission, Canberra.